Guns Up

Naval Action in the Yellow Sea
off Korea, 1950–1953

As the Korean Conflict wore on, frigates, destroyer escorts, cruisers, and battleships of the U.S. Navy, and combatant ships from eight other navies of the United Nations, plus the Republic of Korea Navy, fought a bitter war along the coastlines. Off the west coast of the peninsula, warships operated in treacherous waters of the Yellow Sea, navigating channels between tightly clustered islands close to the mainland. Fluctuating thirty-foot tides, sequentially hid and revealed mud banks, shoals, uncharted rocks, and mines laid in these dangerous coastal waters, covered by enemy shore batteries. The ships toiled to protect both the vital left flank of Allied combat forces ashore, and anti-Communist guerillas operating from nearshore islands to carry out raids behind enemy lines. During bitter armistice talks, these islands became bargaining chips and it was necessary to defend them from enemy shore bombardment and invasion by Chinese and North Korean forces. Through three years of ceaseless warfare, in bone-chilling winters that coated ships with tons of ice, and the sweltering heat of summer that made below-deck areas stifling, Allied sailors stayed the course. One hundred sixty photographs, maps, and diagrams; appendices; a bibliography; and an index to full names, places, and subjects add value to this work.

To the officers and men of the Allied naval ships that plied dangerous Yellow Sea waters off the west coast of Korea

Guns Up

Naval Action in the Yellow Sea
off Korea, 1950–1953

Cdr. David D. Bruhn, USN (Retired)

HERITAGE BOOKS
2021

HERITAGE BOOKS
AN IMPRINT OF HERITAGE BOOKS, INC.

Books, CDs, and more—Worldwide

For our listing of thousands of titles see our website
at
www.HeritageBooks.com

Published 2021 by
HERITAGE BOOKS, INC.
Publishing Division
5810 Ruatan Street
Berwyn Heights, Md. 20740

Copyright © 2021 Cdr. David D. Bruhn, USN (Retired)

All rights reserved. No part of this book may be reproduced or transmitted in any form or by any means, electronic or mechanical, including photocopying, recording or by any information storage and retrieval system without written permission from the author, except for the inclusion of brief quotations in a review.

International Standard Book Number
Paperbound: 978-0-7884-0970-7

Heritage Books by Cdr. David D. Bruhn, USN (Retired)

Battle Stars for the "Cactus Navy":
America's Fishing Vessels and Yachts in World War II

Enemy Waters:
Royal Navy, Royal Canadian Navy, Royal Norwegian Navy,
U.S. Navy, and Other Allied Mine Forces Battling the
Germans and Italians in World War II
Cdr. David D. Bruhn, USN (Retired) and Lt. Cdr. Rob Hoole, RN (Retired)

Eyes of the Fleet:
The U.S. Navy's Seaplane Tenders and Patrol Aircraft in World War II

Gators Offshore and Upriver:
The U.S. Navy's Amphibious Ships and Underwater Demolition Teams,
and Royal Australian Navy Clearance Divers in Vietnam

Guns Up:
Naval Action in the Yellow Sea off Korea, 1950–1953

Home Waters:
Royal Navy, Royal Canadian Navy, and U.S. Navy
Mine Forces Battling U-Boats in World War I
Cdr. David D. Bruhn, USN (Retired) and Lt. Cdr. Rob Hoole, RN (Retired)

Ingram's Fourth Fleet:
U.S. and Royal Navy Operations Against German Runners,
Raiders, and Submarines in the South Atlantic in World War II

MacArthur and Halsey's "Pacific Island Hoppers":
The Forgotten Fleet of World War II

Nightraiders:
U.S. Navy, Royal Navy, Royal Australian Navy, and
Royal Netherlands Navy Mine Forces Battling the
Japanese in the Pacific in World War II
Cdr. David D. Bruhn, USN (Retired) and Lt. Cdr. Rob Hoole, RN (Retired)

On the Gunline:
U.S. Navy and Royal Australian Navy Warships off Vietnam, 1965–1973
Cdr. David D. Bruhn, USN (Retired) and
STGCS Richard S. Mathews, USN (Retired)

Salvation from the Sky: U.S. Navy, Royal Australian Air Force, and
Royal New Zealand Air Force Heroic Air-Sea Rescue in the Pacific in World War II
Cdr. David D. Bruhn, USN (Retired) and Stephen Ekholm

Support for the Fleet:
U.S. Navy and Royal Australian Navy Service
Force Ships That Served in Vietnam, 1965–1973

We Are Sinking, Send Help!:
The U.S. Navy's Tugs and Salvage Ships in the African,
European, and Mediterranean Theaters in World War II

Turn into the Wind
Volume I: US Navy and Royal Navy Light Fleet Aircraft Carriers
in World War II, and Contributions of the British Pacific Fleet

Turn into the Wind
Volume II: US Navy, Royal Navy, Royal Australian Navy, and Royal Canadian Navy
Light Fleet Aircraft Carriers in the Korean War and through End of Service, 1950–1982

Wooden Ships and Iron Men:
The U.S. Navy's Ocean Minesweepers, 1953–1994

Wooden Ships and Iron Men:
The U.S. Navy's Coastal and Motor Minesweepers, 1941–1953

Wooden Ships and Iron Men:
The U.S. Navy's Coastal and Inshore Minesweepers,
and the Minecraft that Served in Vietnam, 1953–1976

Contents

Foreword by Commodore Hector Donohue AM RAN (Rtd)	xxiii
Foreword by Lt Cdr Rob Hoole, RN (Rtd)	xxvii
Foreword by George H. S. Duddy, P.Eng Ret.	xix
Acknowledgements	xxiii
Preface	xxvii
1. Battle of Korea Strait	1
2. Republic of Korea Navy	11
3. Initial United States and Commonwealth Naval Forces	17
4. Allied Nations' Warships	21
5. Naval Blockade and Escort Duties	33
6. Inchon Landings (Operation Chromite)	49
7. Autumn 1950, Wonsan and Chinnampo	61
8. Evacuation of Chinnampo	81
9. Departure/Arrival of Ships in Theater	91
10. Abandonment of Inchon, NW Islands Home to Guerillas	99
11. Blockading Force/Guerilla Cooperation	109
12. Communist Spring Offensive	123
13. Royal New Zealand Navy Action	131
14. Ground War at an Impasse, Armistice Talks Begin	139
15. Enemy Threatens Northwest Islands	149
16. Enemy Island Campaign	157
17. Time Spent "Off the Line"	167
18. Tom Tiddler's Ground	173
19. Ongoing Island Defense	185
20. HMCS *Nootka* Captures North Korean Minelayer	199
21. War Drags On	207
22. Evacuation of West Coast Islands	217
Postscript	225
Appendices	
A. ROKN Ships Acquired Before and During the Korean War	233
B. Battle Stars earned by U.S. Navy Combatant Ships	237
C. Patrol Frigates that served in the Korean War	241
D. U.S. Navy Ships Sunk and Damaged in Action	243
E. Allied Ships' (less USN and ROKN) Deployments	249
F. Cruisers/Destroyers/Frigates assigned to Inchon Landings	255
Bibliography/Notes	259
Index	285
About the Author	299

Photos and Illustrations

Acknowledgements-1: Richard DeRosset	xxiii
Acknowledgements-2: Cdre Hector Donohue, AM RAN (Rtd.)	xxiv
Acknowledgements-3: George H. S. Duddy P.Eng Ret.	xxv
Acknowledgements-4: Lt Cdr Rob Hoole, RN (Rtd.)	xxvi
Preface-1: Flag of the United Nations, New York	xxvii
Preface-2: Battle Honours board of the cruiser HMAS *Hobart*	xxxiii
Preface-3: Ribbons Board of the battleship USS *New Jersey*	xxxiii
Preface-4: Battleship USS *Missouri* firing 16-inch gun salvo	xxxvi
Preface-5: Cruiser USS *Manchester* departing Wonsan Harbor	xxxvi
Preface-6: DDs USS *Wiltsie, Chandler, Hamner,* and *Ozbourn*	xxxvii
Preface-7: Patrol frigate USS *Bayonne* under way	xxxviii
Preface-8: Control vessel USS *PCE(C)-896* replenishing at sea	xxxix
Preface-9: High speed transport USS *Wantuck* under way	xl
Preface-10: UDT members on a Wonsan beach, October 1950	xli
Preface-11: A U.S. Navy fast minesweeper (DMS) under way	xlii
Preface-12: Soviet MiG-15 fighter over North Korea	xliii
Preface-13: USS *Bataan* being screened by HMAS *Bataan*	xlv
Preface-14: HMS *Concord* shelling targets at Chinnampo	lviii
Preface-15: Painting *Battle of Korea Strait* by Richard DeRosset	1
1-1: South Korean sub-chaser *Bak Du San* at Pearl Harbor	1
1-2: Submarine chaser USS *PC-823*	2
1-3: Cargo ship *Alchiba*, circa June 1951	5
2-1: British minesweeper HMS *BYMS-2258*	14
2-2: South Korean sub-chaser *Kum Kang San*	15
2-3: Sub-chasers *Kum Kang San, Chiri San,* and *Sam Kak San*	16
3-1: USN Vice Admirals C. Turner Joy and Arthur D. Struble	17
3-2: Vice Adm. William G. Andrewes, RN	19
4-1: Cruiser HMS *Belfast* coming alongside the carrier USS *Bataan*	23
4-2: British destroyer HMS *Charity*	23
4-3: Australian destroyer HMAS *Tobruk*	24
4-4: New Zealand frigate HMNZS *Hawea*	25
4-5: Royal Canadian Navy destroyer HMCS *Athabaskan*	26
4-6: Rescue operations for survivors of the Thai frigate *Prasae*	27
4-7: Patrol frigates USS *Glendale* and USS *Gallup*	28
4-8: Thai frigate HTMS *Bangpakong*, in the Chow Praya River	28
4-9: Australian frigate HMAS *Condamine* at Williamstown, Victoria	29
4-10: Colombian frigate ARC *Almirante Brion* off Coco Solo	30
4-11: French frigate FS *La Grandiere*	31
5-1: British frigate HMS *Black Swan* at Kure, Japan	36
5-2: British destroyer HMS *Cossack*	39

5-3: Four ships of Destroyer Division 91 at Sasebo, Japan 41
5-4: Drawing of Mokpo, Korea, circa 1951, by Herbert C. Hahn 43
5-5: Canadian destroyer HMCS *Cayuga* off Korea 46
5-6: British frigate HMS *Mounts Bay* at Kure, Japan 46
6-1: Gen. Douglas MacArthur aboard USS *Mount McKinley* 50
6-2: Wolmi-do (Island) at Inchon under bombardment 53
6-3: Capt. Otto H. Becher, RAN, on the bridge of *Warramunga* 57
6-4: Four auxiliary minesweepers of Mine Division 31 58
6-5: USMC 1st Lt. Baldomero Lopez leads Marines over a seawall 59
6-6: Burial at sea for Lt. (jg) David H. Swenson Jr., USN 60
7-1: View of Kunsan, South Korea, looking seaward 61
7-2: Canadian destroyer HMCS *Athabaskan* under way 62
7-3: RCN demolition experts in a dinghy neutralize a mine 64
7-4: Sailors aboard USS *Mockingbird* handling sweep gear 66
7-5: Amphibious shipping anchored in Wonsan's outer harbor 70
7-6: Destroyer USS *Gregory* (DD-802) inside Wonsan Harbor 71
7-7: Rear Admirals Allan E. Smith and Won Il Sohn, ROKN 72
7-8: Sikorsky HO3S-1 helicopter, of HU-1 75
7-9: Royal Air Force Short Sunderland aircraft at Iwakuni, Japan 75
7-10: British light fleet aircraft carrier HMS *Theseus* under way 76
7-11: A LCVP landing craft in the well deck of USS *Catamount* 76
7-12: Seaplane tender USS *Gardiners Bay* off Houghton, Washington 77
7-13: Capt. Stephen Archer, USN 78
8-1: Painting by Bo Hermanson of HMCS *Cayuga* at Chinnampo 81
8-2: Attack transport USS *Bayfield* under way 83
8-3: Destroyer escort USS *Foss* under way 83
8-4: Officers on watch aboard HMAS *Warramunga* 85
9-1: USN and RN flag officers in conference 92
9-2: Vice Adm. Philippe Auboyneau, FN, with American officers 93
9-3: Thai frigate HTMS *Bangpakong* and sub-chaser *Tongpliu* 94
9-4: Thai transport *Sichang* at Singapore 95
9-5: Thai frigate HTMS *Prasae* aground off the east coast of Korea 96
9-6: Four frigates from different navies in Han Estuary, Korea 97
10-1: Capt. Cromwell F. J. Lloyd-Davies, RN 102
10-2: Breech of a naval gun aboard the cruiser HMS *Ceylon* 103
10-3: Heavy cruiser USS *Rochester* under way 105
10-4: Light aircraft carrier USS *Bataan* under way 106
11-1: Three destroyers of British Commonwealth navies 109
11-2: Princess Patricia's Canadian Light Infantry members 110
11-3: Gunner of a 40mm Bofors aboard HMCS *Nootka* 111
11-4: Heavy cruiser USS *St. Paul* under way 112
11-5: Handling of 16-inch projectiles aboard USS *Missouri* 113

11-6: British light cruiser HMS *Belfast* surrounded by sea ice   118
11-7: Ice formed on gun mount of destroyer HMAS *Warramunga*   119
11-8: Rear Adm. Alan Scott-Moncrieff, RN, aboard *Warramunga*   120
12-1: Shore fire control party from the heavy cruiser USS *Toledo*   123
12-2: Gun crew aboard the ROKN patrol frigate *Amnokkang*   125
12-3: Heavy cruiser USS *Toledo* under way   126
12-4: Drawing of northwest Korean coast by Frank Norton   127
12-5: Dutch destroyer HNMS *Van Galen* under way   128
12-6: Two North Korean soldiers near Kaesong   130
13-1: Drawing of HMNZS *Rotoiti* by Frank Norton   131
13-2: Yard minesweeper USS *YMS-413*   132
13-3: New Zealand frigate HMNZS *Rotoiti*   133
13-4: Weary soldiers of 1st Battalion, Royal Australian Regiment   135
13-5: Bren gun practice by an Australian serviceman   137
14-1: UN delegates: Admirals Joy and Burke, and General Hodes   139
14-2: Russian built MiG-15 aircraft at RAAF base at Kimpo   141
14-3: Dock landing ship USS *Whetstone* off Point Loma, California   142
14-4: Wreckage of MiG-15 fighter downed off Korean west coast   143
14-5: Painting of the frigate HMAS *Murchison* by Ken McFadyen   146
14-6: Gun crew manning one of HMAS *Murchison*'s 4-inch guns   148
15-1: Drawing of British cruiser HMS *Ceylon* by Frank Norton   149
15-2: Port side of Canadian destroyer HMCS *Cayuga*   154
16-1: Canadian destroyer HMCS *Athabaskan*   158
16-2: Royal Navy corvette HMS *Sunflower*   160
16-3: British destroyer HMS *Cockade*   161
16-4: UN delegates at Panmunjom, 30 November 1951   163
16-5: Admiral George C. Dyer, USN   165
17-1: Painting of HMAS *Warramunga* by Frank Norton   167
17-2: Repair ship USS *Ajax*, with four destroyers alongside   168
17-3: Drawing of the frigate HMAS *Condamine* by Frank Norton   172
18-1: Ice formed on the Australian frigate HMAS *Condamine*   173
18-2: Vice Adm. Rollo Mainguy aboard HMCS *Athabaskan*   176
18-3: British destroyer HMS *Concord* under way   180
19-1: Canadian destroyer HMCS *Athabaskan* under way   185
19-2: British destroyer HMS *Comus* at Kure, Japan   187
19-3: Marine Corps F4U-4B Corsair aboard USS *Badoeng Strait*   188
19-4: South Korean patrol craft *Kum Kang San* under way   191
19-5: Painting depicting motor and sail junks by Frank Norton   192
19-6: Painting of Wolfpack Headquarters by Frank Norton   193
19-7: Wolfpack guerillas in a boat, and Lieutenant Park, ROKN   194
19-8: RCN Lt. Earl McConechy and Ordnance Lt. Percy Buzza   195
19-9: Comdr. William Landymore and Lt. George MacFarlane   195

19-10: Painting of planes bombing enemy troops by Frank Norton 196
19-11: Escort carrier USS *Sicily* with F4U Corsairs parked aft   197
19-12: British light aircraft carrier HMS *Ocean* at anchor   198
20-1: Canadian Navy destroyer HMCS *Nootka* under way   199
20-2: Dock landing ship USS *Colonial* under way   202
20-3: *Nootka* crewmembers loading one of her 4-inch guns   204
20-4: RCN CPO Joseph Leary supervising the preparation of grog 205
21-1: Drawing of the armistice tent at Panmunjom by Hugh Cabot 207
21-2: Medium landing ship, rocket USS *LSMR-412* under way   208
21-3: Canadian destroyer HMCS *Crusader* under way off Korea   209
21-4: Personnel of ROK Navy aboard a motor torpedo boat   210
21-5: Painting of the frigate HMAS *Condamine* by Frank Norton   212
21-6: Drawing of the Chodo coastline by Frank Norton   213
21-7: British cruiser HMS *Newcastle* at anchor   215
22-1: Korean west coast islands   218
22-2: Vice Admiral Clark, and Korea President Syngman Rhee   220
22-3: Australian frigate HMAS *Culgoa* berthed in port   221
22-4: UN official badge worn by all service and civilian personnel 222
22-5: Friendly-held west coast island, Korea   223
22-6: Private Terry Mahoney guards an entrance to the DMZ   224
Postscript-1: Captain Dacre H. D. Smyth, RAN   225
Postscript-2: Dacre Smyth's depiction of HMAS *Australia*   226
Postscript-3: Lieutenant Dacre Smyth, RAN   229
Postscript-4: Sketch by Smyth of the frigate HMAS *Hawkesbury*   230
Postscript-5: Painting of Commodore Dacre Smyth, AO RAN   232

Maps and Diagrams

Preface-1: Islands off northwest Korea   xlvii
1-1: South Korea, and portion of North Korea   4
5-1: Korea and relatively nearby ports of Sasebo and Kure, Japan   38
5-2: Sea approaches to Inchon, Korea   40
5-3: West coast of Korea between Pyongyang and Inchon   40
7-1: Korea's northeast coast   68
10-1: Area of North Korea surrounding the Hwanghae peninsula   101
14-1: Yonan and Ongjin peninsulas   144
15-1: Key islands located west-southwest of Chinnampo   153
18-1: Hwanghae Province and major offshore islands   176
18-2: Haeju island area, and nearby islands below the 38th Parallel 178

Foreword

David Bruhn's recently published *Turn into the Wind Vol. II*, includes the most conspicuous role undertaken by the U.S. and Commonwealth Navies during the Korean War, that played by the light fleet carriers. His latest book, *Guns Up: Naval Action in the Yellow Sea off Korea, 1950-1953*, covers the operations of U.S., Commonwealth, and other Allied Navy surface combatant ships in the Yellow Sea and complements the description of the carrier operations off the west coast of Korea as described in his earlier book.

Destroyers and frigates assigned to the West Coast Blockading and Escort Force (Task Force 95) screened the U.S., British and Australian carriers, usually stationed about seventy miles west of Inchon. These duties, which involved protection of the carriers and rescue of pilots and aircrew forced to bail out or ditch as a result of combat damage, were more benign than operations closer to shore. Those included shore bombardments and the support of mine countermeasures operations and assistance to guerilla forces.

During the war a total of 76 ships of the Commonwealth Navies and the Fleet auxiliary services served in the war zone for varying periods. These comprised 32 Royal Navy warships (five carriers, six cruisers, seven destroyers, 14 frigates); nine of the Royal Australian Navy (one carrier, four destroyers, four frigates); eight destroyers of the Royal Canadian Navy; six frigates of the Royal New Zealand Navy; two headquarters ships; one hospital ship; 16 Royal Fleet auxiliaries and two merchant fleet auxiliaries.

At the outbreak of hostilities, the *River*-class frigate HMAS *Shoalhaven* was deployed as the Australian naval contingent to the British Commonwealth Occupation Force (BCOF) in Japan, while HMAS *Bataan* was en route to relieve her. Both ships were allocated to United Nations (UN) forces on 29 June 1950 and were immediately included in the Commonwealth Naval Force commanded by Rear Admiral William Andrewes, RN, which was later augmented by ships from Canada, New Zealand, the Netherlands and France. Over the next three years and into the tense post-Armistice period, the RAN maintained two ships on station, as well as deploying an aircraft carrier, a total of some 4,500 personnel. Post-war, the RAN continued Armistice patrols until 1955, including a second tour by the light aircraft carrier HMAS *Sydney*. Fifty-seven officers and men received decorations for their war service.

The four destroyers deployed by Australia during the war were: *Bataan, Warramunga, Anzac* and *Tobruk*; the four frigates were: *Shoalhaven, Murchison, Condamine* and *Culgoa*. The planned deployment period was a 12-month cycle with eight months in Korea and four months on passage. Given the shortages at the time of both ships and manpower, the planned deployment cycle was not achieved. The destroyers averaged some ten months in the war zone, with *Warramunga* spending thirteen months, as a relief ship was not available. The four destroyers each deployed to Korea twice. The frigates each deployed once with *Murchison* and *Condamine* spending nine months in the war zone and *Shoalhaven* and *Culgoa* three and four months respectively.

Three days after the New Zealand government agreed to contribute to the UN naval forces, two frigates sailed from Auckland, arriving at Sasebo on 1 August 1950. A few hours later they were deployed escorting ships between Japan and the South Korean port of Pusan. New Zealand's involvement in the Korean War lasted three years, during which time they maintained two frigates continuously in the war zone. This involved the six *Loch*-class frigates, HMNZ Ships *Pukaki, Tutira, Rotoiti, Taupo, Hawea* and *Kaniere* and about half the manpower of the Royal New Zealand Navy – approximately 1,350 personnel – served in them.

There were no traditional naval battles – control of the seas was firmly held by the UN forces which exploited that control. The RAN destroyers and frigates blockaded the coast, landed and supported raiding parties, supplied isolated UN forces, bombarded coastal targets and escorted larger ships. There was a constant threat from Soviet-built sea mines, especially during the evacuations of Hungnam and Wonsan in December 1950.

Mine clearing was particularly hazardous on the west coast due to the large tidal movements and the tendency of moored mines to 'walk.' Thirteen UN ships were sunk or damaged by Russian-made North Korean mines in 1950. *Warramunga*'s Executive Officer, Lieutenant Commander Geoffrey Gladstone, DSC RAN (later Rear Admiral), was awarded both a bar to his Distinguished Service Cross and the U.S.

Bronze Star for his skill and bravery during his contribution to the minesweeping operations to open the port of Chinnampo in November 1950.

The RAN first engaged the enemy when *Bataan* bombarded a shore battery northwest of Inchon on 1 August 1950. *Warramunga* and *Bataan* took part in the Inchon landing and later returned there to harass advancing Chinese forces after their intervention in the war. On 5 February 1951 *Warramunga* ambushed a North Korean force that, by deception, attempted to lure her inshore.

In September 1951 the *Murchison* operated in the restricted, fast-flowing tidal waters of the Han River, engaging enemy batteries at close range and taking several hits. In addition to combat operations, RAN ships were also involved in humanitarian operations providing food and other supplies to islanders on the west coast who were struggling to survive in the midst of a war zone.

For the last two years of the war, RAN ships in Korean waters continued to protect the islands off the west coast of North Korea that were in South Korean possession.

Korea was the first and last occasion where the UN moved together to wage a large-scale war. From a maritime point of view, everything was in its favour during the Korean war – the physical configuration of the country made it singularly susceptible for the exercise of sea power.

Except for the mining there was practically no enemy opposition at sea. Attacks from the air against ships were negligible and although there was always the possibility of submarine attack, no attacks were detected. Apart from mines, the only serious opposition to ships was from shore batteries during inshore operations and generally speaking, the ships gave more than they received. Korea showed the need for an active and up-to-date minesweeping force, forcibly emphasising the value of amphibious forces and last but not least, showed that the naval gun was not an obsolete weapon.

Perhaps the major lesson from the war, was the certainty that the UN Army could not have existed in Korea without the Navy – the Navy got it there and kept it there.

In *Guns Up*, Commander David Bruhn has provided a comprehensive and interesting description of the many and varied operations undertaken by the UN maritime forces operating in the treacherous Yellow Sea waters in support of the Army ashore. This book is an important contribution to our understanding of what went on in this 'forgotten war.'

Commodore Hector Donohue AM RAN (Rtd)

Foreword

I wasn't even a gleam in my father's eye at the start of the Korean War but I learned of some aspects from two remarkable people in my life. As a recently-arrived English schoolboy in the USA in the mid-1960s, I was picked on by Harry Apetz, my enthusiastically anglophile Social Studies teacher at Bellevue Junior High School near Seattle, and regaled with his tales of serving in the U.S. fleet off Korea. He was effusive in his admiration of the impressive sortie rate achieved by the Royal Navy's smaller but more efficient aircraft carriers.

The other influential figure in my life was Lt Cdr Robbie Robinson MBE RN who was wounded as a junior signalman in the cruiser HMS *Exeter* during the Battle of the River Plate in 1939 and was later 'filleted' (his word) by underwater explosions as a frogman clearing obstructions in the shallows off Dieppe and Normandy. As a Chief Petty Officer, he was a demolitions specialist attached to 41 Independent Commando Royal Marines and participated in the infamously bloody retreat of UN forces from the Chosin Reservoir. He was less than impressed to be told on returning to his young bride in the UK that he was now top of the roster for sea service.

But for the revelations of these two veterans, both of whom have since 'crossed the bar,' I would have known little about the Korean War and might even have believed it was almost exclusively a land campaign involving the odd air battle as depicted in the Commando-type action comics I read as a child. However, after qualifying as an MCDO (Minewarfare & Clearance Diving Officer) in the Royal Navy I found myself lecturing prospective Commanding Officers of warships and submarines about the hazards to be expected from the use of cheap but highly effective mines, and how to deal with them in terms of active and passive mine countermeasures and material, tactical and personal self-protective measures. I cheerfully advised them at the end of each day, "Every ship (or submarine) can be a minesweeper - once!".

One of the most compelling visual aids I used in my lectures was the vivid image of the ROK (Republic of Korea) Navy's minesweeper YMS-516 (formerly the Royal Navy's *BYMS 2148*) being blown asunder by a North Korean mine off Wonsan in October 1950. It was the delayed landing of an amphibious assault force at Wonsan owing to enemy mining, that triggered this famous quotation by Rear Admiral Allen 'Hoke' Smith USN, the Commander of Task Force 95:

> The U.S. Navy has lost control of the seas in Korean waters to a nation without a Navy, using pre-World War I weapons, laid by vessels that were utilized at the time of the birth of Christ.

As ever, David Bruhn has produced a work of historical consequence that sheds fresh light on a little-known area of maritime warfare. He has interwoven his narrative with fascinating snippets that help bring to life the actions of a wide range of units from different allied nations and the bravery of the personnel who manned them. The U.S. Navy, Royal Navy and the navies of the Commonwealth nations of Australia, Canada and New Zealand are given the coverage their predominance deserves but David also pays tribute to the plucky efforts of the embryo ROK Navy which had to overcome enormous odds to field any ships at all. The French and Colombian navies are given their due, too, and there are lessons of interoperability that remain valid to this day. This is manifested in the current deployment of USMC fifth generation F-35B Lightning aircraft, together with those manned by their RN and RAF peers, on board HMS *QUEEN ELIZABETH*, one of the Royal Navy's new 65,000-ton aircraft carriers, during exercises in the North Sea.

Perhaps most significantly of all, David shows what a close-run thing the Korean War was in being waged at all, let alone in its pursuit and outcome. He also explains how the operations of the UN forces ships offshore in their support of the land forces, plus the performance of their aircraft over land and sea, probably tipped the balance between humiliating defeat on the one hand and victory on the other, or at least the uneasy cease-fire that has survived for the past 70 years. Long may it continue.

Rob Hoole
Vice Chairman & Webmaster
RN Minewarfare & Clearance Diving Officers' Association
13 October 2020

Foreword

For Canadian destroyers, Korea was a peculiar war. Having come out of World War Two as an effective anti-submarine navy, our ships suddenly found themselves engaged in the monotony of interminable carrier screening missions and hazardous blockade or island defence patrols. They were often called upon to complete these operations in the limited visibility of snow squalls and biting winds, among the rocky shoals and mud flats of the Korean west coast. It is no small wonder our sailors were happy when the so-called United Nations' 'police action' was over and our ships could return to Canada.

—Fred R. Fowlow, former Supply Officer aboard
HMCS *Athabaskan* during the Korean War.[1]

Royal Canadian Navy Ensign 1911-1965

United Nations Flag

The headlines of Victoria's *The Daily Colonist* of July 6, 1950, read "Ontario Escorts Three Destroyers Headed for Pearl Harbor." As a boy of 10 with a keen interest in the Royal Canadian Navy, I distinctively remember that day over 70 years ago when the cruiser HMCS *Ontario* lead the tribal destroyers HMCS *Cayuga* and *Athabaskan* and the V-class with the tribal name HMCS *Sioux* out of Esquimalt Harbor. *Ontario* would soon return to base after refueling the others at sea but they would continue to a final location not formally announced to their crews, but strongly rumored to be Korea.

Canada was proud to serve in Korea with its Commonwealth, U.S., and ROK Partners. Significantly, for the first time, combat operations, although basically commanded by the U.S. Military, were under aegis of the United Nations. This is an organization since its inception that Canada as a middle power has always firmly supported. It is noteworthy, as detailed in this book, ten nations participated in the naval actions in Korea—a consensus in military action that perhaps may never be repeated.

It may have been one of the UN's finest actions in defending a small independent nation from unwanted attack and invasion – a nation that until 1991 was not even one of its members. In that year she and her fervent northern enemy were both granted membership. Sadly, their bitter conflict has not yet been resolved but the UN action has permitted those in the south to survive and prosper in freedom while those in the north starve beside their missiles and bombs.

Fortunately, in spite of conflicting Cold War demands in Europe and in the midst of a restructuring of her naval forces back to a specialist ant-submarine force, Canada was able to quickly respond to the Korean conflagration by the rapid deployment of destroyers from her west coast. Decisive and rapid government action to support the war was important, because arrival of the destroyers in theater required a voyage of thousands of miles across the Pacific from their base at Esquimalt on southern Vancouver Island. Mobilization and demobilization voyages for later deployments of HMCS *Iroquois*, *Huron*, *Haida*, and *Nootka* based at Halifax Nova Scotia were even longer, the first three completed two circumnavigations of the globe while *Nootka* completed one.

Canadian postal stamp tribute to *Tribal*-class destroyers, circa 1942

As a boy and youth, who had a dream of a naval career, I did gain some personal knowledge of some of the RCN vessels which served in Korea. My dad and myself often trolled for salmon in Parry Bay near Victoria on Vancouver Island and I remember passing closely to HMCS *Cayuga* and *Athabaskan* my favorite vessels. Our friends and good next-door neighbors were Angus and Nan Rankin. Commander Rankin was the captain of HMCS *Sioux* on her final and Canada's last tour of duty in Korea, which lasted until 1955.

Although my desire to serve in the RCN was not realized, I have also felt a kinship with the sea. The Royal Canadian Navy and other Commonwealth navies are patterned on, and closely linked to the Royal Navy. It's been said, at least in years past, that one would be hard pressed to find an establishment in England to which one goes to "bend an elbow" without a painting of a ship or the Royal Navy. The memorizing of naval ships' names there is a passion similar to the American one of recalling names of their movie stars – it was one I inherited from my dad.

In Canada, like in Britain, there is much interest in naval matters. I was aware of the deployments of our ships to Korea, their distinguished participation in "train busting" on the east coast, and even the strange involvement of Ferdinand Waldo ("Fred") Demara, the Great Imposter, who untrained in medicine, performed lifesaving surgery on HMCS *Athabaskan*. However, many of their fine contributions on the west coast service were new to me, prior to my involvement with and review of the *Guns Up* manuscript.

The RCN destroyers, like those of the other Commonwealth navies, alternated carrier screening duties, with patrol, shore bombardment, and island protection duties. Using its boat, carrying a demolition party, HMCS *Athabaskan* rendered enemy moored mines, exposed at low tide, safe by blowing them up with attached charges. HMCS *Nootka* also helped combat the threat of Soviet mines emplaced by North Korea by capturing one of their minelayers. This stealthy vessel, used in offshore channels, was a large junk, whose superstructure had been removed, leaving a mere 18 inches of freeboard to lessen the chance of visual detection. To ensure quietness when engaged in nefarious activities, the craft had been propelled by oarsmen, in lieu of a noisy engine disclosing its presence to passing UN patrol vessels.

NAVAL GUNFIRE ACTION

> *It is safe to say that throughout the duration of the Korean War, Canadian warships sent off thousands of rounds of 4-inch and 40 mm rounds into enemy troop concentrations, at moving trains, garrisons, the inshore 'gunboat navy' comprised of mine laying junks, guerilla troop transports and shore-based gun emplacements. In one forty-day patrol period HMCS* Athabaskan *expended 1,050 rounds of 4-inch and 590 rounds of 40 mm.*
>
> — Fred R. Fowlow, former Supply Officer aboard
> HMCS *Athabaskan* during the Korean War.[2]

Facing almost no naval threat from the North, other than its liberal use of mines and infrequent aircraft attacks, RCN destroyers focused on adversaries ashore. Enemy shore artillery fire, bent on the destruction of ships, added to the dangers posed by the mines, and geography, encountered while protecting the flanks of ground forces ashore. Canadians are proficient in operating in high northern latitudes, offering bitter cold, pack ice, and navigation in constricted waters. Extreme tidal changes, fast running currents, and other impediments associated with combat duty in the Yellow Sea, such as exposed mudflats and barely concealed rocks were, however, a new experience.

The book's title, *Guns Up*, reflects the preponderance and also frequency of naval gunfire operations in Korea: against enemy targets; in support of friendly ground forces, and anti-Communist guerilla raids ashore; and in ongoing efforts to interdict enemy logistics. The latter activity frequently resulted in duels with shore artillery, particularly on the east coast. Many ships experienced hits or near misses. HMCS *Iroquois* suffered the only RCN members killed in action in Korea, when "B" gun deck was hit by an enemy round. Lt. Comdr. John Quinn and Able Seaman Elburne Baikie died instantly. Able Seaman Wallis Burden died several hours later of his wounds. Ten other crewmembers suffered wounds caused by shrapnel fragments and the blast.

LEST WE FORGET

RCN Destroyers that served in the Korean War

Pacific Coast Command	Ship Class	Atlantic Coast Command	Ship Class
HMCS *Athabaskan* (DDE219)	Tribal	HMCS *Haida* (DDE215)	Tribal
HMCS *Cayuga* (DDE218)	Tribal	HMCS *Huron* (DDE216)	Tribal
HMCS *Crusader* (DDE228)	*C*-class	HMCS *Iroquois* (DDE217)	Tribal
HMCS *Sioux* (DDE225)	*V*-class	HMCS *Nootka* (DDE213)	Tribal

Summary of Personal Honours and Awards

Commonwealth Awards	No.	U.S. Awards to RCN	No.
Distinguished Service Order (DSO)	1	Legion of Merit	7
Order of the British Empire (OBE)	3	Distinguished Flying Cross	1
Distinguished Service Cross (DSC)	9	Bronze Star Medal	1
Bar to the DSC (Second Award)	1		
Distinguished Service Medal (DSM)	2		
British Empire Medal (BEM)	4		
Mention in Despatches (MID)	32[3]		

George H. S. Duddy, P.Eng Ret.

Acknowledgements

Masterful, and prolific maritime and aviation artist Richard DeRosset has completed over a thousand paintings to date in his illustrious art career, influenced by much time spent at sea as a sailor in the U.S. Navy, deckhand on a commercial fishing vessel, and finally as master of the small tanker *Pacific Trojan*. His painting *Battle of Korea Strait*, the cover art for this book should be of great pride to every South Korean, particularly the officers and crew of the patrol craft ROKN *Bak Du San* (PC-701). In a sea battle on the first night of the Korean War, these men defeated a larger armed vessel loaded with North Korean soldiers intending to capture the port of Pusan. This action kept the port in friendly hands, enabling the Allies to surge troops and war materiel to the beleaguered Republic of Korea.

Photo Acknowledgements-1

Richard DeRosset at work in his art studio in Lemon Grove, California. Courtesy of Victoria Maidhof

I am particularly grateful to Commodore Hector Donohue, AM RAN (Retired), for his considerable assistance with this and previous books. In addition to providing the perspective of a Royal Australian Navy flag officer in forewords and postscripts, he has also generously shared material from the many published books and articles he has authored.

Donohue began his career in the RAN, in 1955, as a seaman officer and subsequently sub-specialized as a clearance diver and torpedo and anti-submarine officer. His service in the RAN included command of the destroyer escort HMAS *Yarra* and the guided missile frigate HMAS *Darwin*. Ashore, he held a number of senior positions in Defence policy and Force development prior to retirement in mid-1991.

Photo Acknowledgements-2

Hector Donohue while in command of HMAS *Darwin*.
Courtesy of Commodore Donohue, AM RAN (Rtd).

This book would not have been possible without the considerable involvement of Canadian George Duddy, a retired Professional Engineer with a keen interest in maritime subjects. An expert on the maritime history of western Canada and the Arctic, he is also quite interested in naval history. This is likely due to his lineage. His father, as a schoolboy, took photographs of surrendered German battleships in the Firth of Forth near the end of World War I; his great grandfather was a pioneering Leith steamship owner and his great great grandfather, as Master in both sail and steam, ended his career as marine superintendent for the Leith, Hull & Hamburg Steam Packet Co. One of his other relatives, Midshipman Percival George, tragically fell to his death from the mast of a Royal Navy ship during the age of sail. His dirk is on display in a museum in South Africa.

Duddy offered many suggestions for improvement during his technical review and editing of this book, as he has done for others; made available material related to the Royal Canadian Navy; and provided valuable insight and perspective in his foreword related to the subject matter of the book.

Photo Acknowledgements-3

George Duddy in Glacier Bay aboard the MS *Volendam* during an Alaskan cruise in 2019. A contributor and U.S. Military & Naval Vessel Correspondent for Nauticapedia.ca Project, he is sporting an organization ballcap.
Courtesy of George Duddy

Rob Hoole co-authored with me the *Home Waters/Nightraiders/Enemy Waters* trilogy of mine warfare books and, for this book, reviewed the Royal Navy-related material, and graciously penned a foreword. Rob is a former Royal Navy mine clearance diving officer and commanding officer of the mine countermeasures vessel HMS *Berkeley*. An acknowledged expert on mine warfare, he is a long-standing member of the Ton Class Association and a regular contributor to its publications. Hoole is also founding Vice Chairman and Webmaster of the Royal Naval Minewarfare & Clearance Diving Officers' Association, and holds key positions in related organisations.

Rob spearheaded successful efforts to establish a memorial at Gunwharf Quays in Portsmouth, UK, in remembrance of the tens of thousands of service personnel who passed through the gates of the training establishment HMS Vernon. These personnel included the mine warfare and diving specialists, mine designers, minefield planners, bomb & mine disposal personnel and the crews of the minelayers, minesweepers, and minehunters who were trained or based there.

Photo Acknowledgements-4

Rob Hoole standing in front of the Vernon Mine Warfare & Division Monument at the Gunwharf Quays in Portsmouth, UK. Courtesy of Rob Hoole

Canadian John M. MacFarlane allowed use of material from his article "Some Notes on HMCS *Iroquois* in the Korean War," as well as a photograph of *Iroquois* conducting shore bombardment of enemy targets in North Korea. Formerly the Director and Curator of the Maritime Museum of British Columbia in Victoria, MacFarlane is now the Curator of the website, The Nauticapedia.ca, which shares a vast resource related to British Columbia's nautical history, and has two extensive searchable databases of ships and people. He has written many articles and books on this theme. His latest book, *Around the World in a Dugout Canoe: The Untold Story of Captain John Voss and the Tilikum*, co-authored with Lynn J. Salmon, was recently published.

Finally, much thanks to my stalwart editor, Lynn Marie Tosello, who has joined me in my quest to honor generally unsung sailors and their ships. Her scouring and polishing of hundreds of pages of text, while gently, in most cases, identifying overlooked omissions, or the need to better explain nautical terms, adds much to these works. She also brings to her work, humor in addition to eloquence and prose.

Preface

One lesson that I taught my crew which I had learned from World War II, was that if you wanted to make good gunners out of mediocre gunners, simply take them under enemy fire.

—Comdr. James A. Dare, commanding officer of the destroyer USS *Douglas H. Fox* (DD-779).[1]

I have been much impressed by the way in which the navies of so many nations are co-operating in the Korean War. In spite of the differences in language and customs, warships of different nations are operating as one against the common enemy.

—Words of praise from Adm. Sir Roderick McGregor, GCB, DSO Royal Navy First Sea Lord and Chief of Naval Staff, following an inspection tour to Korea.[2]

I have only the highest praise for the manner in which our allies contributed to the war effort of the UN Navy. Their co-operation was all that could be desired and they performed every task assigned them, no matter how difficult, with zeal and ability that always evoked my admiration.

—Vice Adm. Charles Turner Joy, USN, commander, Naval Forces Far East.[3]

Photo Preface-1

The Flag of the United Nations, New York, 1950.
Naval History and Heritage Command photograph #NH 97167

Well before daylight on 25 June 1950, six North Korean infantry divisions and three Border Constabulary Brigades, supported by Soviet-made T-34 tanks, heavy artillery, and the North Korean Air Force, swarmed south across the 38th parallel, touching off the Korean War. The Soviet-supported North Korean Army advanced rapidly overland against poorly trained and ill-equipped Republic of Korea (ROK) forces while, along South Korea's east coast, a Border Constabulary Brigade carried out amphibious landings. On 26 June, two more North Korean divisions moved south across the 38th parallel.[4]

On 30 June, President Harry S. Truman committed U.S. troops to enforce a UN Security Council resolution, asking member nations to provide necessary assistance to South Korea. Twenty-one nations committed themselves to support South Korea which was not yet a member nation; sixteen supplied fighting units and five sent military hospitals and field ambulances.[5]

The first U.S. troops to fight in defense of South Korea engaged North Korean military forces at Osan, 30 miles south of Seoul, with tragic results. Task Force SMITH (540 men of the U.S. Army 24th Infantry Division, quickly dispatched from occupational duties in Japan) was crushed, on 5 July, by the North Korean 4th Division. In response to this devastating defeat, a UN Command was created, on 7 July, under the command of Gen. Douglas MacArthur.[6]

The following week, Lt. Gen. Walton Walker, the commander of the U.S. Eighth Army, was assigned responsibility for all ground operations in Korea. Following heavy losses at the hands of the North Korean 3rd and 4th Divisions, American and ROK troops were forced to retreat further south to the Naktong River, the last natural physical barrier to Pusan.[7]

REPEAT OF DUNKIRK SEEMINGLY POSSIBLE

On 4 August, driven south by the North Korean Army, three American and five South Korean army divisions were forced to make a final stand along the 145-mile-long Pusan perimeter, the last UN toehold in Korea. The situation was so grim that six days earlier, General Walker had issued a "Stand or Die" order to the troops. The alternative was to be pushed to the sea and annihilated.

When the first elements of the 1st Provisional Marine Brigade arrived at Pusan from San Diego, on 2 August, fear abounded in the city. It was a similar condition to one that had existed in World War II, when German troops drove the British Expeditionary Force to the water's edge at Dunkirk, France. Pinned against the sea, the British and other Allied soldiers of the Force would have been destroyed, were it not for a heroic, and successful evacuation effort ordered by British Prime Minister Winston Churchill. In addition to efforts by the Royal Navy, and Royal Air Force, hundreds of private craft made the Channel crossing to save "the sons of England."[8]

ASSISTANCE FROM THE SEA REQUIRED

With the rapidly deteriorating situation on the peninsula where, it appeared, the Eighth Army might be driven into the sea, it seemed only the UN naval forces had the power to reverse the situation. Facing the distinct possibility of imminent military defeat on land, Vice Adm. C. Turner Joy, USN, commander, Naval Forces Far East (ComNavFE), ordered Task Force 95 (the Blockade and Escort Force) to bombard the important railway lines running the length of North Korea's eastern seaboard. Down this coastal route, fed by six rail lines from Manchuria and the connecting Trans-Siberian line, flowed all the war material for the two major approaches to the Pusan perimeter.[9]

ORGANIZATION OF UN NAVAL FORCES

On 25 June 1950, at the outset of the Korean War, Vice Adm. Turner Joy had only one cruiser, four destroyers, four amphibious ships, one submarine, ten minesweepers, and the Australian frigate HMAS *Shoalhaven* under his command. This situation changed within forty-eight hours, as Adm. Arthur W. Radford, commander in chief, U.S. Pacific Fleet, transferred operational control of the Seventh Fleet (essentially all of the USN ships in the western Pacific) to Vice Admiral Joy. The following month, all United Nations naval forces committed to assisting South Korea were also placed under the operational control of commander, Naval Forces Far East (Joy). This mix of ships was quickly organized into four separate task forces:

- TF 77 (Carrier Strike Force)
- TF 95 (Blockade and Escort Force)
- TF 96 (Naval Forces, Japan)
- TF 90 (Far East Amphibious Force)[10]

During the interim period, before the assignment of all UN naval forces to ComNavFE, other steps were also being taken to increase naval strength in the theater. On 27 June, as directed by commander in chief, Pacific Fleet (CincPacFlt), Vice Adm. Struble, at Buckner Bay, Okinawa, reported for duty to commander in chief, Far East, Gen. Douglas MacArthur. One day later, the British Admiralty placed Royal Naval units in Japanese waters at the disposal of Vice Admiral Joy, who requested that the British ships rendezvous at Buckner Bay. Requiring additional ships, CincPacFlt formed Task Force Yoke under Rear Adm. Walter F. Boone, USN. Serving as commander, Cruiser Destroyer Group, Seventh Fleet, Boone's job was to scrounge up as many ships as possible on the U.S. West Coast and at Pearl Harbor, for the Korean campaign.[11]

On 5 July, ComNavFE (Vice Admiral Joy) implemented President Truman's order for a blockade of the Korean coast and, five days later, he directed that the blockade be extended northward to include the ports of Wonsan (east coast) and Chinnampo (west coast). That same day, Adm. Forrest P. Sherman, chief of Naval Operations (CNO), directed CincPacFlt to sail Task Force Yoke when ready. On 11 July, the CNO authorized activation of ships from the Reserve Fleet. As part of this process, a Patrol Frigate Activation program was begun at Yokosuka, on 15 July, to put some of the ships lent the Soviets after World War II, and later returned to U.S. custody at Yokosuka, back in service.[12]

On 24 July, ComNavFE established Escort Element (CTE 96.50) under Capt. Alan D. H. Jay, RN—consisting of the frigates HMS *Black Swan*, HMS *Hart*, and HMAS *Shoalhaven*. An escort force was needed to shepherd transport ships and merchantmen being surged to Pusan, on Korea's southern coast, to deliver desperately needed combat troops, war material, and supplies.[13]

While bolstering support for United Nations forces in Korea, ComNavFE concurrently needed to disrupt supply lines running down Korea's east coast to enemy troops advancing southward. The major part of this effort was shore bombardment by ships offshore, and strikes by carrier aircraft. Additionally, on 27 July, Admiral Joy directed harassing and demolition raids by Task Force 90, utilizing UDT (Underwater Demolition Team) and Marine reconnaissance personnel against selected North Korean east coast military objectives.[14]

UNITED NATIONS NAVAL FORCES

The naval forces of the U.S., Britain, Australia, and Republic of Korea in the theater of war, were soon augmented by additional warships of these countries, as well as ones dispatched by Canada, New Zealand, France, the Netherlands, and later Thailand and Columbia.

Surface combatant ships of the UN Naval Forces were assigned to support the Carrier Strike Force (TF 77); the Blockade and Escort Force (TF 95); Naval Forces, Japan (TF 96); and the Far East Amphibious Force (TF 90). More about these duties in later pages.

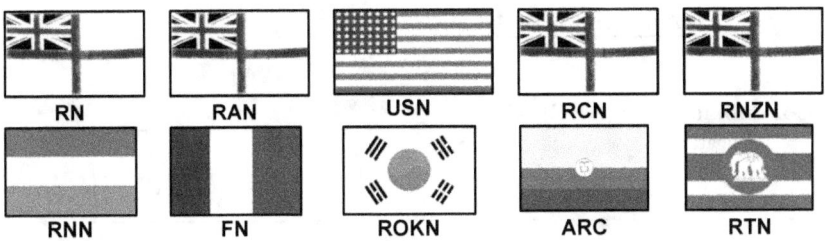

The United States Navy and Royal Navy were able to commit the largest number of powerful ships to the UN Naval Force; smaller navies made contributions relative to their size. A summary of the UN Naval Force surface combatant ships follows (326 vessels total). The fire power of the Royal Navy ships was much greater than those of the Republic of Korea, which had more vessels. The ROK Navy aggressively employed its smaller ships and craft in offshore and, more often, critical and difficult inshore combat operations.

U.S. Navy (USN) – 211

4	Battleships	5	Fast Transports (converted destroyers)
9	Cruisers	13	Patrol Frigates
172	Destroyers	4	Patrol Escorts
4	Destroyer Minesweepers		

Royal Navy (RN) – 37

6	Cruisers	14	Frigates
7	Destroyers		

Republic of Korea Navy (ROKN) – 54

5	Patrol Frigates	2	Gunboats
6	Submarine Chasers	2	Patrol Gun Boats
5	Patrol Craft Sweepers	19	Yard Minesweepers (ex-YMS/BYMS)
4	Motor Torpedo Boats	11	ex-Japanese Minesweepers (JMS)

Royal Australian Navy (RAN) – 8

4	Destroyers	4	Frigates

Royal Canadian Navy (RCN) – 8

8	Destroyers

Royal New Zealand Navy (RNZ) – 6

6	Frigates

Royal Netherlands Navy (RNN) – 4

3	Destroyers	1	Frigate

Royal Thai Navy (RTN) – 4

4	Frigates

Colombian Navy (ARC) – 3

3	Frigates

French Navy (FN) – 1

1	Frigate

Information about, and the identities of the Republic of Korea Navy ships may be found in Chapter 2 and in Appendix A. Chapter 4 is devoted to the ships of the other countries, with the exception of those of the United States.

BATTLE HONOURS AND BATTLE STARS

The ships of the Royal Navy and the other Commonwealth Nations, with successful war service, earned "Battle Honors KOREA." Those of the United States that garnered one or more battle stars were able to garnish the ribbon boards on their deckhouses. (It is unknown to the author, whether any of the other nations—Columbia, France, Netherlands, the Republic of Korea, and Thailand—authorized similar campaign awards for eligible ships of their navies.)

Because His/Her Majesty's warships (whether RN, or units of other Commonwealth nations) do not carry Army regimental colours, battle honours are instead displayed on a battle honour board. Traditionally teak, this solid wooden board is mounted on the ship's superstructure, carved with the ship's badge and scrolls naming the ship and the associated honours. The board is either completely unpainted, or with the lettering painted gold. To pay tribute to past ships of the same name, their honours are displayed as well. Battle Honours (which date back to the year 1588, when 'ARMADA 1588' was authorized) are awarded for six types of action:

- Fleet or Squadron Actions
- Single-ship or Boat Service Actions
- Major Bombardments
- Combined Operations
- Campaign Awards
- Area Awards[15]

Photo Preface-2

Battle Honours board of the Royal Australian Navy light cruiser HMAS *Hobart* (D63), depicting the ship's battle honours and badge. Since she was the first HMAS ship with this name, her honours resulted from her war service only.
Australian War Memorial photograph 300798

Photo Preface-3

Ribbons board on the bridge wing of the battleship USS *New Jersey* (BB-62). Her Korean Service Medal ribbon, with four battle stars, is the one located on the right in the fourth row from the top.
Courtesy of John Werda

U.S. Navy ships and units had to meet one of the following criteria to be considered to have participated in combat operations:
- Engaged the enemy
- Participated in ground action
- Engaged in aerial flights over enemy territory
- Took part in shore bombardment, minesweeping, or amphibious assault

- Engaged in or launched commando-type raids or other operations behind enemy lines
- Engaged in redeployment under enemy fire
- Engaged in blockade of Korean waters
- Operated as part of carrier task groups from which offensive air strikes were launched
- Were part of mobile logistic support forces in combat areas[16]

The officers and men of U.S. Navy ships awarded battle stars were eligible to affix one or more stars to their Korean Service medal or ribbon worn on uniform blouses, and their ships to display same on their bridge wing ribbon boards.

KOREAN WAR ENGAGEMENTS

Battle stars were authorized for the following ten engagements. The catch was, only one star was authorized for each engagement, no matter how many qualifying actions by a particular ship. Thus, the number of stars earned by a USN ship was related both to qualifying actions, and to time spent in theater.

1. North Korean aggression
2. Communist China aggression
3. Inchon landing
4. 1st U.N. counteroffensive
5. Communist China, spring offensive
6. United Nations summer-fall offensive
7. 2nd Korean winter
8. Korean defense, summer-fall (1952)
9. 3rd Korean winter
10. Korea, summer-fall (1953)[17]

Two hundred eleven U.S. Navy surface combatant ships collectively garnered 814 battle stars—on average, a little less than four apiece. A summary of those earned by each ship may be found in Appendix B. The top twenty-three ships, identified below, each earned 7-9 stars. While appreciating their efforts, it is important to note that Battle Honours KOREA awarded to Commonwealth ships, might encompass several engagements, and/or combat actions.

U.S. NAVY SURFACE COMBATANTS AWARDED THE MOST BATTLE STARS IN THE KOREAN WAR

★★★★★★★★★★

USS *Manchester* (CL-83)
USS *Chevalier* (DDR-805)

USS *Theodore E. Chandler* (DD-717)
USS *Wiltsie* (DD-716)

★★★★★★★★★

USS *Endicott* (DMS-35)
USS *George K. Mackenzie* (DD-836)
USS *Hamner* (DD-718)
USS *Hanson* (DDR-832)

USS *Saint Paul* (CA 73)
USS *Southerland* (DDR-743)
USS *Taussig* (DD-746)

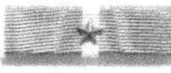

USS *Brinkley Bass* (DD-987)
USS *Duncan* (DDR-874)
USS *Eversole* (DD-789)
USS *Gloucester* (PF-22)
USS *Henry W. Tucker* (DDR-875)
USS *James E. Keyes* (DD-787)

USS *John A. Bole* (DD-755)
USS *John W. Thomason* (DD-760)
USS *Leonard F. Mason* (DD-852)
USS *Lofberg* (DD-759)
USS *Rupertus* (DD-851)
USS *Thompson* (DMS-38)[18]

DD: destroyer
DDR: radar picket destroyer
DMS: high-speed minesweeper
    (converted destroyer)
CA: heavy cruiser
CL: light cruiser
PF: patrol frigate

Korean Service Medal, and Ribbon with one battle star

Before delving into the text of the book, an overview of a few ship types of the U.S. Navy sent to Korea (which had the largest and most diverse group of warships) might be useful. Their introduction, to a degree, also serves as a primer on the duties in which they engaged. No explanations accompany the photographs of the battleship, cruiser, and destroyers, only captions.

Photo Preface-4

Battleship USS *Missouri* (BB-63) firing a salvo from her No. 2 turret, while bombarding Chongjin, North Korea, October 1950.
Naval History and Heritage Command photograph 80-G-421049

Photo Preface-5

A sailor aboard USS *Manchester* (CL-83) observes the enemy coastline as the cruiser, shore bombardment completed, departs Wonsan Harbor.
Naval History and Heritage Command photograph #NH 97187

Photo Preface-6

Destroyers USS *Wiltsie* (DD-716), USS *Theodore E. Chandler* (DD-717), USS *Hamner* (DD-718), and USS *Ozbourn* (DD-846) anchored at Pearl Harbor, circa 1949. These four ships collectively earned 31 battle stars in the Korean War.
Navy History and Heritage Command photograph #NH 96199

PATROL FRIGATE ACTIVATION PROGRAM

Thirteen patrol frigates served in the Korean War, a relatively modest portion of the UN forces. Brought out of "mothballs" at Yokosuka, they were pressed into service by the U.S. Navy when need for warships, whatever size, was greatest in the early dark days of the war. Three of these ships were subsequently transferred to the Republic of Korea, and two each to Thailand and Columbia. These seven ships continued their combat service in the Korean War, as part of the UN naval forces, but under the flags of their respective nations.

In WWII, American and Canadian shipbuilders constructed ninety-six, 304-foot patrol frigates, of a design (S2-S2-AQ1) that was based on the Royal Navy's *River*-class corvette. Two of these ships (of 301 feet in length) were built by Canadian Vickers. Of this fleet, 37 were almost immediately transferred to the U.S. Dept. of Commerce for service as weather ships. Another 21 (built in Providence, Rhode Island, by the Walsh-Kaiser Co.) were re-designated *Colony*-class frigates and delivered to the RN at the builder's yard. Of the remaining 38 USN patrol frigates, when the war ended, 28 were transferred to the USSR; six went to other nations, and four were scrapped.[19]

Thirteen of the ships transferred to the Soviet Union at the end of WWII, and subsequently returned to U.S. custody, served in the Korean War. They had been decommissioned in 1945 at Cold Bay, Alaska, and loaned to the USSR under the Lend-Lease Program, then returned to the U.S. Navy, in late 1949 and early 1950, and laid up in the Pacific Reserve Fleet, Yokosuka. On 15 July 1950, following the outbreak of war in Korea, the USN began a patrol frigate (PF) activation program at Yokosuka.[20]

Photo Preface-7

USS *Bayonne* (PF-21) under way in Far Eastern waters, circa 1950-1953.
Naval History and Heritage Command photograph #NH 56362

The patrol frigates taken out of the Reserve Fleet for war service are identified in Appendix C. The six PFs not transferred to Allied navies remained a part of the U.S. naval forces, until transferred to the Japanese Maritime Self-Defense Force in 1953, near war's end or following it.[21]

UNNAMED AMPHIBIOUS CONTROL VESSELS

The smallest U.S. Navy surface combatants in Korea were four unnamed 180-foot ships. The USS *PCE(C)-822*, *PCE(C)-886*, *PCE(C)-896*, and *PCE(C)-898* were remnants of a class of 68 patrol craft escort vessels built and delivered to the U.S. Navy and an additional 17 delivered under the Lend-Lease Program to Allies during World War II.[22]

Propelled by diesel engines to a modest top speed of 15 knots, and originally armed with a dual-purpose 3"/50 gun, three 40mm guns, five 20mm guns, two depth charge tracks, and ten K-Guns, PCEs were an inexpensive substitute for larger and more valuable destroyers and destroyer escorts. The PCE was designed for general escort work; some were converted to PCE(R), rescue escorts, while others were converted to amphibious control vessels, PCE(C).[23]

The PCE(C)s that served in the Korean War were employed in this role, leading UN landing forces to hostile beaches, and employing their guns as necessary in protection of assault forces.[24]

Photo Preface-8

USS *PCE(C)-896* replenishing at sea.
U.S. Navy photo from *All Hands* magazine, January 1956

HIGH-SPEED TRANSPORTS

Four 306-foot fast transports comprising Transport Division 111 based at Yokosuka—*Horace A. Bass* (APD-124), *Begor* (APD-127), *Diachenko* (APD-123), and *Wantuck* (APD-125)—served in Korea. Laid down in their builders' yards in World War II as *Rudderow*-class destroyer escorts, they had been modified to serve as troop transports, capable of carrying and launching amphibious forces.[25]

The resultant high-speed transport combined the hull and (reduced) armament of a warship, with the superstructure of a troop transport. To accomplish this dual role, only the forward mount was fitted of the two 5-inch guns that a destroyer escort (DE) would carry. Elimination of the after mount allowed a substantial length of the main deck to be enclosed to house 160 troops. Another change involved adding a cargo hold aft, with a crane capable of handling light vehicles and equipment. Finally, the three-tube centerline torpedo mount found on the destroyer escort class, was removed to free up space for port and starboard side boat stations capable of launching and recovering four thirty-six-foot landing boats.[26]

Photo Preface-9

High speed transport USS *Wantuck* (APD-125) under way.
Naval History and Heritage Command photograph #L45-301.02.01

Although unable to carry out a DE's anti-submarine mission, having no torpedo tubes, the high speed transports (APDs) still had fire power necessary for the missions in which they engaged. In addition to its main armament (one 5-inch gun), each high-speed transport boasted three twin 40mm gun mounts, aimed either optically or through the ship's fire control radar system, and additional smaller 20mm mounts aimed optically by their gunners.[27]

APD EMPLOYMENT IN KOREA

> *The sailors became very protective, possessive even, of "their" raiders. To the ship's crew it became a matter of pride, if not outright honor, that the ship not let the raiders down when the going got tough on the North Korean coastline. This principle was irrevocable, whether the raiders be American, Korean, or British Royal Marine Commandos, all of whom entered the unique world of APD operations.*
>
> —Crewmember of a high-speed transport, describing the special bond between ship's company and the raiders they carried to perform special operations ashore.[28]

Photo Preface-10

Lt. Dan F. Chandler briefs Underwater Demolition Team members on a Wonsan beach, 26 October 1950. The frogmen were there to destroy a North Korean minefield. National Archives photograph #80-G-421429

APDs operating singly or in pairs (with commander, Transport Division 111, normally embarked), engaged in nearshore combat operations. These missions involved carrying American frogmen (UDT members), CIA-led Korean guerillas, and British Royal Marine Commandos to hostile shores, and bringing them safely back to base. The frogmen used a variety of demolitions in their work, but the standard charge was the Mark-135 Demolition Pack, which contained twenty pounds of C-3 plastic explosive. Individual weaponry taken by the UDT members (frogmen) behind enemy lines included submachine guns, pistols, and knives for the close-quarters combat that characterized most raiding missions.²⁹

Korean guerillas inserted across North Korean beaches were sent ashore in small teams, at night, to conduct limited reconnaissance missions, establish Escape and Evasion networks, and to collect local intelligence, particularly on the railway system. For the most part, they were North Korean civilians, screened and recruited by the CIA from among the numerous refugee camps around Pusan. Those selected were taken to the CIAs guerilla training base on the small island of Yong-do, located ninety miles south-southwest of Pusan. There, the Koreans were put through an accelerated training program by a small number of U.S. military personnel "on loan" to the Agency from their respective services.³⁰

High-speed transports and embarked frogmen of UDTs 1 and 3 also engaged in mine clearance. Aided by Russian advisors, North Korea mined the major harbors within its control, including the east coast ports of Wonsan and Hungnam; and Chinnampo, the west coast port of entry for the North Korean capital city of Pyongyang.³¹

FAST MINESWEEPERS

Photo Preface-11

A U.S. Navy Fast Minesweeper (DMS) with the number "18" painted on her hull, while representing the fictitious minesweeper USS *Caine*, during the filming of the movie *The Caine Mutiny*, circa 1954, staring Humphrey Bogart, Jose Ferrer, Van Johnson, Fred MacMurray, and Robert Francis. The ship is probably the USS *Doyle* or *Thompson*. Naval History and Heritage Command photograph #NH 84548

A small group of former U.S. destroyers also participated in the Korean War as fast minesweepers: *Carmick* (DMS-33), *Doyle* (DMS-34), *Endicott* (DMS-35), and *Thompson* (DMS-38). Built as 348-foot *Bristol*-class destroyers, they were later re-equipped and designated fast minesweepers. *Thompson* served as a destroyer after World War II, but she later joined her three sister ships to comprise Mine Squadron One.[32]

Ordered to Korea to help combat the mine threat, *Thompson* and *Carmick* departed San Diego, on 4 October 1950, and arrived at Sasebo on the 21st. Lengthy mine clearance operations at Chinnampo ensued, for which the two fast minesweepers received the Navy Unit Commendation for their actions during the period 28 October-20 November 1950. Opening the port proved fortuitous, in respect to forthcoming events. Following the intervention of Communist China in the Korean War, in late November 1950, its 250,000 ground forces threatened to cut off and destroy UN units operating in the mountains of North Korea. To prevent that possible catastrophe, on 9 December, Gen. Douglas MacArthur ordered evacuation by sea of the U.S. Tenth Corps.[33]

These evacuations were carried out by naval forces at Hungnam and Chinnampo, on the east and west coasts, respectively. At Chinnampo, while UN warships shelled advancing Communist troops, *Thompson* escorted troopships out of the harbor in dense fog and through treacherous tidal currents to assist in the evacuation.[34]

ENEMY MIGS, MINES, AND SHORE ARTILLERY

Photo Preface-12

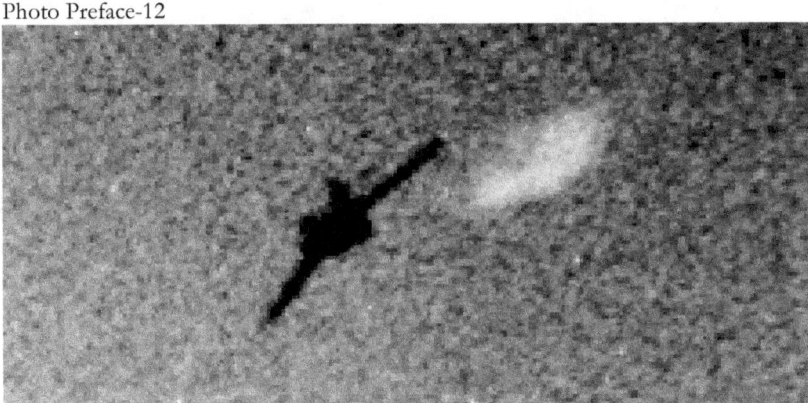

Soviet MiG-15 jet fighter during an air battle in which it was shot down over North Korea by F9F-2 Panther fighters from USS *Leyte*, 23 November 1950.
National Archives photograph #80-G-424091

Simultaneous to the entry of Chinese ground forces into the war, came a new threat when Russian pilots, dressed in Chinese uniforms, entered the skies over the Yalu. The appearance of the Soviet MiG-15 came as a huge surprise to the UN coalition forces. On 9 November 1950, MiG-15 planes attacked F9F Panther jet aircraft from USS *Philippine Sea* in the first engagement between planes of this type. On 18 November, eight F9Fs of the Fast Carrier Task Force engaged eight to ten MiG-15s, shooting down one and damaging another five.[35]

The Korean War was peculiar in that there was, initially, no effective enemy opposition on the seas. North Korea, unable to contest UN control of the seas by air, ship, or submarine, did so by other means—sea mines set adrift, and laid in harbor approaches, and ports; and shore artillery firing on Allied ships. Cruisers and destroyers regularly closed the coast when functioning as seagoing heavy artillery. This was necessary to provide naval gunfire support to Allied ground troops ashore; and when shelling railway bridges, tunnels, and trains (and other targets) to sever logistic support to the enemy.[36]

Confronting mines and engaging in duels with shore guns was dangerous work. Five U.S. Navy ships were sunk by mines (four of them minesweepers) during the war, and others were damaged. Between 13 September 1950 (when three U.S. Navy destroyers were hit by enemy gunfire at Inchon while conducting pre-landing shore bombardment) and 11 July 1953, there were 87 incidents in which USN ships were damaged in action. In the last action, the heavy cruiser *Saint Paul* (CA-73) suffered severe under water damage from a 76-90mm hit from a shore battery at Wonsan. Summary information about USN ship damage and personnel casualties may be found in Appendix D.[37]

FOCUS OF THE BOOK

Guns Up is devoted to operations of United States, Commonwealth, other Allied Navy, and Republic of Korea warships in the Yellow Sea. The book complements descriptions of carrier operations off the west coast of Korea, found in my book, *Turn into the Wind, Vol. II, U.S. Navy, Royal Navy, Royal Australian Navy, and Royal Canadian Navy Light Fleet Aircraft Carriers in the Korean War and through end of service, 1950-1982*. As previously noted, while larger U.S. Navy fleet carriers mainly kept to the deeper Sea of Japan off Korea's east coast, their svelte sisters—the light aircraft carriers USS *Bataan*, HMS *Glory*, HMS *Ocean*, HMS *Theseus*, HMS *Triumph*, and HMAS *Sydney*—were consigned to the Yellow Sea. (Please note that the Republic of Korea objects to the name "Sea of Japan" for the body of water which is bordered by Japan, Korea, and Russia. It instead favors "East Sea." Sea of Japan is used herein, because it is most familiar to American readers and those of Commonwealth and other countries.)

HMS *Triumph*
Battle Honour
Korea 1950

HMS *Theseus*
Battle Honour
Korea 1950–51

HMAS *Sydney*
Battle Honours
Korea 1951–52
Malaysia 1964
Vietnam 1965–72

HMS *Glory*
Battle Honour
Korea 1951–53

HMS *Ocean*
Battle Honour
Korea 1952–53

USS *Bataan*
7 Battle Stars

Photo Preface-13

Light fleet carrier USS *Bataan* off the west coast of Korea, being screened by the Australian destroyer HMAS *Bataan* (foreground) and other surface combatants. A rare photo showing two ships named for the same battle. (RAN)

The four British and single Australian light carriers operated off Korea's west coast, alternating duty with the American light aircraft carrier *Bataan* (CVL-29), or escort carrier USS *Badoeng Strait* (CVE-116), USS *Barioko* (CVE-115), USS *Rendova* (CVE-114), and USS *Sicily* (CVE-118). At the end of an operation period, when relieved by its successor, the carrier on station proceeded to Sasebo, Kure, or Kobe, Japan, for re-supply, maintenance, and crew rest.

Destroyers, frigates, and patrol frigates assigned to the West Coast Blockading and Escort Force (TF 95) screened the British, American, and Australian carriers, usually stationed about sixty miles southwest of Haeju. These duties, involving protection of the carriers, and retrieval from the sea of pilots and aircrews forced to bail out or ditch as a result of combat damage, were more benign than those that took them closer to shore—those being shore bombardment, and the support of mine clearance operations and guerilla forces. Destroyers and frigates, operating in shoal waters, provided protection to west coast islands on which guerilla forces were based, and from which they launched operations behind enemy lines.[38]

CRITICALLY IMPORTANT NEARSHORE ISLANDS

A long string of islands originating off northwest Korea, extended down around the peninsula to Pusan in the southeast. Some of these islands were of great tactical importance to Allied forces. Friendly, guerilla-held islands were ideally suited for radar units and signal intercept stations, and also served as bases for other elements. They offered safe haven for helicopter teams and boat crews dedicated to rescuing downed airmen. Islands behind enemy lines, in particular, served as springboards for guerilla actions and agent insertions.[39]

Additionally, guerillas occupied key terrain that controlled several Yellow Sea choke points. Their control of these islands limited enemy movements around the mouth of the Yalu River, into the port cities of Chinnampo and Haeju, and within the important Han River Estuary. Allied control of the sea, increased with the guerilla assistance, forced all supply support for the enemy front lines to move overland or by rail, making them vulnerable to air attacks and naval gunfire.[40]

This changed, in November 1951, when Communist forces overwhelmed 1,000 defenders on the western islands near the mouth of the Yalu and seized them. In recognition of the vulnerability of critical islands to amphibious assault, additional support for the Republic of Korea Marines and guerillas holding the outposts was soon forthcoming. On 6 January 1952, commander, Blockade and Escort Force became responsible for the defense of all islands north of the 38th Parallel along both coasts. Total responsibility for the sea, air, and land elements of northern island defense was now vested, for the first time, in a single commander.[41]

The following map shows the location of the headquarters of U.S. Marine Col. James T. Wilbur, commander, West Coast Island Defense Element (CTE 95.15). Paengyong-do Island, on which he was based, lay just below the 38th parallel.[42]

Map Preface-1

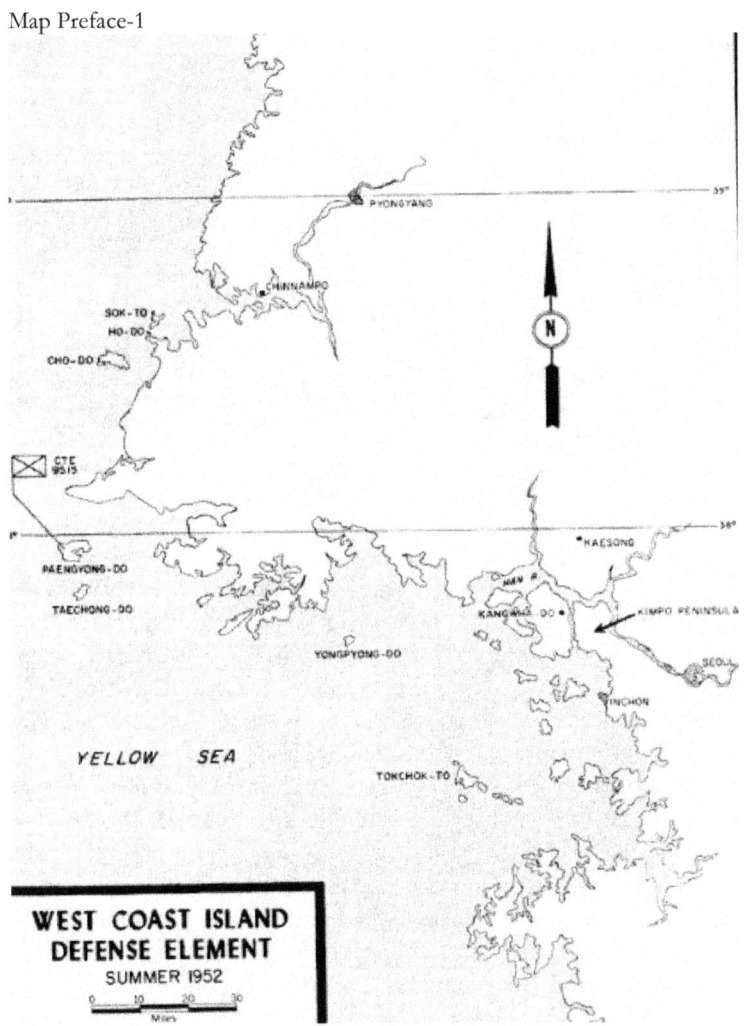

Islands off northwest Korea, above and below the 38th parallel
The Sea Services in the Korean War, 1950-1953[34]

SHIP DUTY AND CONDITIONS IN THE YELLOW SEA

Photo Preface-14

British destroyer HMS *Concord* (D63), photographed from HMAS *Bataan* (D191), at the entrance to Chinnampo on the west coast of North Korea. *Concord*, surrounded by pack ice, was carrying out a shore bombardment of the port installations of the town. Australian War Memorial photograph 042342

While the piston-engine aircraft aboard the carriers attacked enemy supply lines, fortifications, and troop positions, enemy MiG jet aircraft were a constant threat, some flown by Russians. Carrier air (and destroyers and frigates, operating independently from aircraft carriers) also provided protection to friendly islands, from which guerillas mounted offensive operations behind enemy lines.

Off the Korean west coast—ragged and heavily indented with numerous small islands—winters were cold, with occasional gales and blinding snow squalls, and the sweltering summers were hot and humid, with heavy rains and fog. Combat operations in coastal waters were made difficult by these conditions attendant to the geography and climate, along with extreme tidal changes, and associated shoal water, exposed mudflats, and fast-running currents that made navigation among the tiny islands that dotted the coastline, particularly perilous.

Ships carrying out gunfire assignments (shore bombardment of enemy targets or support of guerilla activities ashore) were often at navigational peril, as they made their way to points that would bring targets under their guns. Lt. Rodney Agar, RN, gunnery officer aboard the destroyer HMS *Concord*, described these operations and of the wearing out of gun barrels over time, as a result of the ship's volume of fire:

> On the west coast of Korea as one of the bombarding ships we would creep in at night among the islands, or by day carry out air-spotted shoots, in support of the Army's flanks. We eventually wore out our gun barrels and when the Gunner passed the 'plug bore' tests the plug just fell out of each barrel! [This refers to a tampion, a plug that goes into the muzzle of the gun when not in use to keep sea spray out of the barrel. In the USN, it is pronounced tom-kin.][43]

As previously described, conditions aboard ship were particularly arduous in the winter. In at least one ship, the food and water making capability was also marginal, as recalled by Comdr. Daniel Herlihy, RNZN (Retired). Herlihy had joined the Royal New Zealand Navy at age 17 in 1949 as a Seaman Boy, and was serving aboard the frigate HMNZS *Taupo* at the outbreak of the Korean War:

> Onboard conditions were harsh. We had inadequate wet and cold weather gear – a British issue duffel coat that the wind whipped up, sea boots and stockings and a pair of long johns. We were grateful for the NZ Patriotic Board's present of woolen gloves and balaclava. Heating was poor and we were sleeping down below the waterline and it was cold. The inside doors of the upper deck would ice over in winter.
>
> Food was out of the deep fryer but I was young and had no complaints. Washing conditions were tricky especially with water rationing in the summer despite the distillation units onboard ship.[44]

With this introduction in our wake, readers may now (vicariously) stand out to sea with U.S., Commonwealth, and warships of other Allied nations off the west coast of Korea.

| Preface

Photo Preface-15

Battle of Korea Strait by Richard DeRosset depicts the sea battle, on the night of 25 June 1950, between the South Korean sub-chaser *Bak Du San* (PC-701) and a freighter, likely the SS *Kimball R. Smith*, in the Korea Strait between Pusan and the Japanese Island of Tsushima. The patrol craft was an ex-U.S. Merchant Marine Academy training ship; the freighter, a former U.S. Army coastal cargo ship on loan to the government of South Korea. *Kimball R. Smith* was serving as a training vessel, when her South Korean crew mutinied and defected to the north. On the night of the first day of the Korean War, the freighter, which was packed full of North Korean soldiers intending to capture the port of Pusan, was sunk due to heroic actions by the captain and crew of *Bak Du San*. Had access to Pusan been blocked by a Communist presence, Allied ships could not have delivered desperately needed troop reinforcements, and the war might have been quickly lost.

1

Battle of Korea Strait

Photo 1-1

South Korean sub-chaser *Bak Du San* (PC-701) receiving a 3"/50 gun mount at Pearl Harbor, in March 1950. Six .50 caliber machine guns would complete her armament. Naval History and Heritage Command photograph #NH 97002

On the night of 25-26 June 1950, the sub-chaser *Bak Du San* (PC-701) was patrolling the Korea Strait between Pusan and the Japanese Island of Tsushima (see Map 1-1). Based at Chinhae, on the southeast coast of South Korea, she had a crew of about sixty officers and men, and was the most capable ship in the Republic of Korea Navy (ROKN).[1]

Photo 1-2

Submarine chaser USS *PC-823*, location and date unknown.
Historical Collections of the Great Lakes

Bak Du San was the former sub-chaser USS *PC-823*, built by the Leathem D. Smith Shipbuilding Corp., Sturgeon Bay, Wisconsin. Commissioned on 24 July 1944, she served in the western Atlantic during World War II. The 173-foot steel-hulled ship had a 23-foot beam, a 450-ton displacement, and a 10-foot draft. Two 2,560 bhp Hooven-Owen-Rentschler RB-99DA diesel engines were capable of propelling her to a maximum speed of 20 knots.[2]

On 11 February 1946, as part of U.S. Navy post-war downsizing, she had been decommissioned and transferred to the Maritime Commission. On 18 May 1948, she was transferred to the U.S. Merchant Marine Academy (sited on the Hudson River at Kings Point, New York) for use as a training ship and named *Ensign Whitehead*. Demilitarization of the warship, sometime before her handoff to "Kings Point," would have included removal of her armament:

- one 3"/50 dual purpose gun mount
- one single-barrel 40mm gun mount
- three 20mm guns
- two rocket launchers
- four depth charge projectors, and two depth charge tracks[3]

The Republic of Korea Navy (ROKN) purchased the former *PC-823*, its first significant warship, in autumn 1949, using funds partially donated by officers and men of the service. Senior officers down to cadets, contributed 5-10 percent of their salary. Some midshipmen sold scrap metal to earn additional contribution money, and their wives helped by earning funds taking in laundry and sewing. The ROKN accumulated $15,000, but not nearly enough to buy a naval vessel. Learning of the ROKN's laudable efforts, the South Korean government provided an additional $45,000 to make up the balance of the $60,000 purchase price.[4]

After acquiring the *Ensign Whitehead*, fifteen ROKN naval officers spent two months in the United States preparing her for the voyage to South Korea. She was in such bad shape, that initially only one of her engines worked. The renamed and renumbered *Bak Du San* (PC-701) (some references cite *Pak Tu San* as her name) sailed from the port of New York for Pearl Harbor where she was rearmed with one 3-inch gun. A hundred rounds of ammunition for it were purchased in Guam. The pride of the Republic of Korea Navy arrived at Chinhae Naval Base South Korea, on 10 April 1950, barely two months before the 25 June 1950 outbreak of the Korean War. With only 100 rounds of ammunition available, her sailors could only practice aiming the main gun, not firing it.[5]

Before dawn on 26 June, while patrolling the Korea Strait, *Bak Du San* sighted an unidentified ship. The mysterious vessel had neither a naval ensign nor a flag aloft; there also was no name on her transom, or numbers on the sides to disclose nationality or identity. Attentive watch standers aboard PC-701, aided by a waxing gibbous (three-quarters illuminated) moon in the night sky, had uncloaked an infiltrator from the north trying to skirt the patrol craft and reach Pusan Harbor.[6]

Bak Du San challenged by flashing light and, receiving no response, turned her searchlight on the intruder. The light revealed a freighter with an estimated six hundred soldiers crowded on her decks. Heavy machine guns, mounted aft on the ship, quickly opened fire. Rounds striking the patrol craft's bridge killed the helmsman and seriously wounded the officer of the deck. *Bak Du San* returned fire with her single 3-inch gun, and machine guns. Gun crews at exposed battle stations, poured round after round into the enemy transport during an ensuing running battle, until finally sinking it in waters between Pusan and Tsushima Island.[7]

Map 1-1

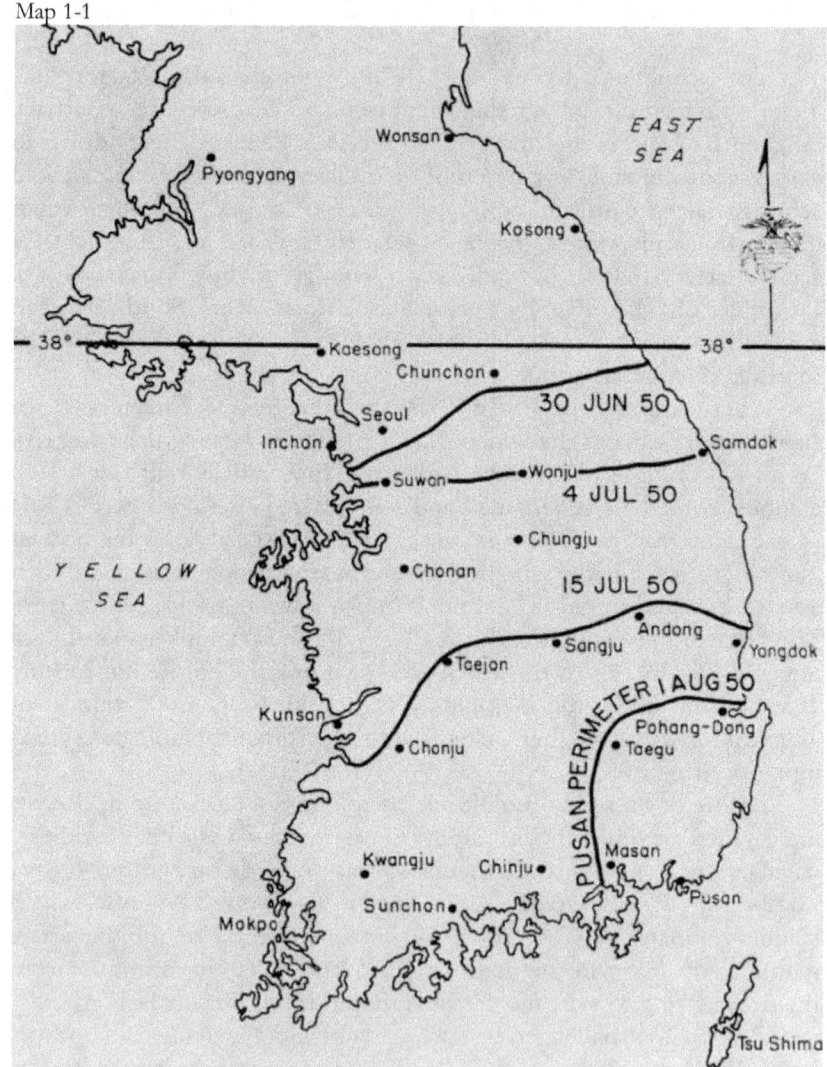

South Korea, and portion of North Korea, above the 38th parallel (North Korean Army gains, 30 June–1 August 1950)
Map from *The Inchon-Seoul Operation, U.S. Marine Operations in Korea, 1950–53, Vol. II*

To increase the accuracy and penetration of his ship's gun rounds, Comdr. Nam Choi Yong, ROKN, had closed to within 400 meters of the freighter. As the enemy ship slipped into the deep, the surface of the water was suddenly covered with thrashing formerly embarked troops. As explained by her crew, recognizing the threat the North Korean soldiers still presented, sailors aboard *Bak Du San* used M1 Garand-rifle fire to prevent the enemy from attempting to swim to, and board the patrol craft.[8]

NORTH KOREAN ENEMY SHIP

Photo 1-3

Cargo ship *Alchiba* (AK-261), circa June 1951. Built by Ingalls Shipbuilding, Decatur, Alabama, she was a sister ship of SS *Kimball R. Smith*.
Naval History and Heritage Command photograph #NH 84173

The vessel North Korea had employed as a troop transport was likely the SS *Kimball R. Smith*, laid down as *Northern Captain* by the Ingalls Shipbuilding Corp., Decatur, Alabama. Completed in June 1945, she was delivered to the U.S. Army as one of fifty-nine N3-S-A2 coastal cargo ships built by the Maritime Commission. Nineteen were operated in Southwest Pacific Area, as part of the U.S. Army's permanent local fleet through the end of World War II and into the 1950s. The first of these coasters arrived in theater on 5 September 1944 and the last in December 1945.[9]

The steel-hulled vessels were single-deck coasters, characterized by a vertical stem and a cruiser stern. *Kimball R. Smith*, like her sisters, was 259 feet in length with a 2,760-ton displacement. Propelled by oil-fired boilers, a steam reciprocating engine and single propeller, she could make an economical 11 knots—with clean hull.[10]

A few of these small U.S. Army cargo ships (not including *Kimball R. Smith*) were acquired by the U.S. Navy for service as post-war, non-commissioned auxiliaries. *Kimball R. Smith*, and at least six ex-U.S. Navy cargo ships—*Alchiba* (AK-261), *Algorab* (AK-262), *Aquarius* (AK-263), *Centaurus* (AK-264), *Cepheus* (AK-265) and *Serpens* (AK-266)—were leased to South Korea. *Alchiba* and *Kimball R. Smith* were products of the same builder's yard. *Alchiba* was a Navy cargo ship in World War II, then saw service with the U.S. Army Transportation Corps. as USAT *Charles S. Winsor*. Reacquired by the Navy, on 12 June 1951, she was renamed *Alchiba*. South Korea returned her to U.S. custody, in January 1960, and she was struck and sold for scrapping.[11]

SOUTH KOREAN CREW SAILS SS *KIMBALL R. SMITH* TO NORTH KOREA

> *I am instructed by my Government to bring to your Excellency's attention the matter of the U.S. merchant vessel,* Kimball R. Smith. *The vessel is of the N-3 type.... The* Kimball R. Smith *left Pusan, Korea on September 20 with a cargo of salt destined for Kunsan, Korea. It is on loan from the U.S. Government to the Government of the Republic of Korea.... The two officials of the United States Economic Cooperative Administration sailed with the vessel as advisors to the Korean crew. The officials are Alfred T. Meschter ... and Albert Willis...*
>
> *On September 24 the Pyongyang radio station in Korea announced the arrival of the* Kimball R. Smith *at the port of Chinnampo on September 22.... In view of the fact that no further word has been received, it would be appreciated if the Soviet Government would lend assistance in ascertaining the exact location of the vessel, the welfare of the two American officials, and would facilitate the prompt departure of the officials and the vessel in order that they may proceed to the port of original destination.*
>
> —Excerpts from a note delivered, on 1 October 1949, by Ambassador Alan G. Kirk at Moscow, to the Foreign Office of the USSR.[12]

Making surplus U.S. Navy-acquired cargo ships available to South Korea in the early 1950s followed earlier practices. On 29 May 1949, the cargo ship SS *Kimball R. Smith* (registered in Mobile, Alabama) arrived in South Korea. She was one of five shallow-hulled coastal transports on loan from the U.S. government to South Korea as part of the American Mission in Korea (AMIK).[13]

Kimball R. Smith was serving as a training vessel, when her South Korean crew mutinied and defected to the north—sailing the cargo ship to Chinnampo, a port near Pyongyang. Two American merchant mariners, Captain Alfred T. Meschter and Chief Engineer Albert C. Willis, were aboard to provide instruction for the crew. The United States, which did not recognize the North Korean Communist regime, made two requests to Russia for aid in locating the two Americans, without satisfactory results. Acting Soviet Foreign Minister Andrei Gromyko replied, on 14 October, that it was a matter solely for the North Korean government.[14]

Meschter (age 28) and Willis (40) were detained in North Korea for 81 days until their release at the 38th parallel to John J. Muccio, the first American Ambassador to South Korea.[15]

HEROIC ACTIONS BY *BAK DU SAN*'S CREW MAY HAVE CHANGED COURSE OF FLEDGLING WAR

> *As the sun disappeared behind the rugged Korean skyline and darkness enshrouded the land, the Brigade sailed into the port of Pusan on 2 August [1950]. There we found thousands of Korean refugees in the crowded port in a state of bedlam, gloom and defeatism, a foreboding characteristic of the early days of the Korean War. We encountered badly frightened military and civilian personnel who demonstrated a behavioral attitude of "doomsday is coming."*
>
> —Col. Robert D. Taplett, USMC (Ret.), commander of the 3rd Battalion, Fifth Marines, describing in *Dark Horse Six*, the conditions facing the 1st Provisional Marine Brigade upon its arrival at Pusan, Korea.[16]

While the actions described in the preceding sections constituted a major victory for South Korea at the start of the Korean War, its greater significance can only be fully appreciated by considering it in the context of the overall disastrous state of the war that was unfolding at the time.

On 25 June 1950, well before daylight on a Sunday morning, 135,000 North Korean troops had launched an invasion of South Korea in an all-out surprise attack. Following 45 minutes of artillery bombardment, six North Korean infantry divisions and three Border Constabulary Brigades, supported by Soviet-made T-34 tanks, heavy artillery, and the North Korean Air Force, swarmed across the 38th parallel. Aimed at reuniting the country under Communist rule from the North, the hostile act touched off the Korean War.[17]

The Soviet-supported North Korean Army advanced rapidly overland against poorly trained and ill-equipped Republic of Korea forces while, along South Korea's east coast, a Border Constabulary Brigade carried out amphibious landings at Kangnung and Samchok. On 26 June, following a demand by the UN Security Council that North Korea terminate its attack and return to its borders, two more North Korean divisions moved south across the 38th parallel. The following day, the UN Security Council approved a resolution, introduced by the U.S., asking member nations to provide necessary assistance to South Korea. The measure passed because the Soviet Union representative was absent and thus unable to veto it.[18]

Three days later, President Harry S. Truman committed U.S. troops to enforce the UN demand and some 16 countries also committed and subsequently sent military forces to South Korea under the UN flag. Many other countries contributed equipment, supplies, and other support. North Korea's principal supporters were the Soviet Union, which supplied it with arms and military advisors, and Communist China, which later poured masses of troops into the conflict.[19]

The first U.S. troops to fight in defense of South Korea engaged North Korean military forces at Osan, 30 miles south of Seoul, with tragic results. Task Force SMITH, which had been quickly dispatched from occupational duties in Japan to fight a delaying action against overwhelming odds, was crushed, on 5 July, by the North Korean 4th Division. The U.S. unit, comprised of only 540 men of the 24th Infantry Division, had been misinformed about its objective, was undertrained for its mission, and was poorly equipped with castoff gear from WWII.[20]

In response to this devastating defeat, the United Nations Command was created, on 7 July, under the command of Gen. Douglas MacArthur. The following week, Lt. Gen. Walton Walker, the commander of the U.S. Eighth Army, was assigned responsibility for all ground operations in Korea. Following heavy losses at the hands of the North Korean 3rd and 4th Divisions, American and ROK troops were forced to retreat further south to the Naktong River, the last natural physical barrier to Pusan.[21]

At month's end, additional U.S. troops arrived at Pusan, first the 9th Infantry Regiment and the 2nd Infantry Division and, two days later, the Marines. The 1st Provisional Marine Brigade, from Camp Pendleton, was attached to the already in-place U.S. Army 25th Infantry Division and moved jointly forward as reserves at Masan, located north-northwest of Pusan. By 4 August, however, despite the addition of these troops, Walker had been pushed farther seaward toward Pusan, the last allied stronghold on the Peninsula. Walker then ordered the defense of a 140-mile arched boundary around the city, thereafter termed the "Pusan Perimeter."[22]

Kwon Yule-jung, chief director of the Busan Regional Office of Patriots and Veterans Affairs of South Korea, described in an article, titled "Great Battle of Korea," the significance of the little-known, Battle of Korea Strait, which took place, on 26 June 1944, to the future of South Korea:

> We should not forget that without the Battle of the Korea Strait and the decisive victory by the southern sailors of the warship *Baekdu Mountain* [*Bak Du San*], the war would have likely ended in less than one month and in the North's favor. Then we all might have lived under the last three generations of North Korean dictatorship.
>
> If the North's specialized forces had landed in Busan [then Pusan], they could have taken over the city which was not ready for any battle, and began the push north, maybe up to Daegu [then Taegu], without meeting any resistance as their forces descended from the north. The war might have ended right there.[23]

2

Republic of Korea Navy

Perhaps the most aggressive and effective, if smallest, member of the South Korean armed services during the first year of the Korean War was the Republic of Korea Navy (ROKN).

—Edward J. Marolda, former Acting Director of U.S. Naval History and Senior Historian of the Navy, in an article in 2017.[1]

The Republic of Korea Navy (ROKN) was established on 15 August 1948, but its roots date back to the creation of the Korean Maritime Affairs Association, on 11 November 1945, by a merchant mariner, Sohn Won Il. This action followed the liberation of the Korean Peninsula from Japan, on 15 August 1945. The Association evolved into the Marine Defense Group, on 11 November, and later became the Korean Coast Guard, which was formed in Chinhae.[2]

In 1946, a contingent of U.S. Coast Guard officers, led by Capt. George McCabe, arrived in South Korea to help organize, supervise, and train the sea service. These Coast Guardsmen worked closely with their Korean counterparts for nearly two years, during which time McCabe jointly commanded the service alongside Lt. Comdr. Sohn Won Il.[3]

The Korean Coast Guard, in carrying out its area responsibilities, utilized mainly ex-Imperial Japanese Navy (IJN) and U.S. Navy minecraft as patrol vessels. Vessels acquired by the Korean Coast Guard prior to its transition to ROKN status are identified in the table.

LCI 351-class Infantry Landing Craft - 6

Ship	Formerly	Comm.
Seoul (LCI-101)	ex-USS *LCI(M)-594*	29 Oct 46
Jinju (LCI-102)	ex-USS *LCI(G)-516*	11 Nov 46
Chuncheon (LCI-103)	ex-USS *LCI(L)-773*	3 Jan 47
Cheongju (LCI-104)	ex-USS *LCI(G)-453*	3 Jan 47
Cheongjin (LCI-105)	ex-USS *LCI(M)-1056*	22 Jan 47
Jinnampo (LCI-106)	ex-USS *LCI(G)-442*	22 Jan 47

Patrol Gunboat - 1

Ship	Formerly	Comm.
Chungmugong I (PG-313)	Japanese rescue ship; construction unfinished at war's end	7 Feb 47

Yard Minesweepers (YMS) /British Yard Minesweepers (BYMS) - 18

Ship	Formerly	Comm.
Kang Jim (YMS-501)	ex-USS *YMS-354*	28 Apr 47
Kyong Chu (YMS-502)	ex-USS *YMS-358*	28 Apr 47
Kwang Chu (YMS-503)	ex-USS *YMS-413*	28 Apr 47
Gaeseong (YMS-504)	ex-HMS *BYMS-2006*	2 Jun 47
Kim Hae (YMS-505)	ex-USS YMS-356	2 Jun 47
Ganggye (YMS-506)	ex-USS *YMS-392*	2 Jun 47
Kang Nung (YMS-507)	ex-USS *YMS-463*	21 Jun 47
Kang Wha (YMS-508)	ex-USS *YMS-245*	11 Nov 47
Gapyeong (YMS-509)	ex-USS *YMS-220*	11 Nov 47
Ganggyeong (YMS-510)	ex-USS *YMS-330*	11 Nov 47
Kaya San (YMS-511)	ex-USS *YMS-423*	11 Nov 47
Guwolsan (YMS-512)	ex-USS *YMS-323*	11 Nov 47
Kim Chon (YMS-513)	ex-HMS *BYMS-2258*	11 Nov 47
Kil Chu (YMS-514)	ex-HMS *BYMS-2005*	21 Sep 47
Gyeongsan (YMS-515)	ex-HMS *BYMS-2018*	21 Sep 47
Gongju (YMS-516)	ex-HMS *BYMS-2148*	20 Feb 47
Gowan (YMS-517)	ex-USS *YMS-473*	11 Nov 47
Yong Kung (YMS-518)	ex-HMS *BYMS-2008*	Jul 47

Tug Minelayers (ex-Japanese Mine-Planting Vessels) - 11

Ship	Formerly	Comm.
Daejeon (JMS-301)	No. 1313	11 Nov 46
Tongyeong (JMS-302)	No. 1314	11 Nov 46
Daegu (JMS-303)	No. 1372	11 Nov 46
Taebaeksan (JMS-304)	No. 1373	11 Nov 46
Dumangang (JMS-305)	No. 1121	9 Jan 47
Danyang (JMS-306)	No. 1009	8 Apr 47
Dancheon (JMS-307)	No. 1269	3 Oct 47
Toseong (JMS-308)	*Daiichikadogawamaru*	3 Oct 47
Daedonggang (JMS-309)	No. 1008	3 Oct 47
Deokcheon (JMS-310)	No. 1217	3 Oct 47
Tongcheon (JMS-311)	No. 1216	9 Jan 47

Fuel Oil Barge - 1

Ship	Formerly	Comm.
Guryong	USS *YO-118*	24 Dec 46

With the establishment of the Republic of Korea (South Korean) government, on 15 August 1948, the Korean Coast Guard was renamed the Republic of Korea Navy, with Sohn serving as its first chief of Naval Operations. With transition to Navy status, came additional ships, with many more following during the Korean War. These included patrol frigates, sub-chasers, motor torpedo boats, patrol craft sweepers, gunboats, and additional amphibious vessels (tank landing ships and landing support ships). A complete list of ROKN ships acquired before and during the war (78 total), may be found in Appendix A.[4]

PHILIPPINE NAVY (PN) ASSISTANCE

Members of the Philippine Navy played a pivotal role in the formation of the Korean Coast Guard, by bringing the first ships of the KCG to Korea from Subic Bay Naval Base. In August 1947, Lieutenant Senior Grade Ramon A. Alcaraz, PN, was designated officer in charge of a group tasked with sailing former U.S. and British motor minesweepers (YMS/BYMS) to their ports of destination. The backbone of the South Korean Fleet was to operate from Chinhae, Pusan, and Seoul. The ROKN's bases would also include Mokho, Mukpo, Kusan, Pohang, and Inchon.[5]

The Philippines did not send combatant ships to Korea, but its navy did play a supporting role in the war. Its army did take part. On 7 September 1950, President Elpidio Rivera Quirino announced the deployment of Filipino troops to South Korea. Only twelve days later, on 19 September 1950, the 10th Battalion Combat Team (BCT) arrived at Pusan, after a four-day voyage from Manila; carried aboard the Military Sea Transportation Service ship USNS *Sgt. Sylvester J. Antolak* (T-AP-192). The transport was escorted by Philippine patrol craft RPS *Negros Oriental* (ex-USS *PC-1563*) and RPS *Capiz* (ex-USS *PC-1564*) from the vicinity of Corregidor Island to the outskirts of the South China Sea.[6]

The battalion was the first of five BCTs deployed—the 10th, 20th, 19th, 14th, and 2nd—each serving in sequence for about a year in Korea, with the last departing for Manila in 1955. Combat service support operations of the Philippine Navy actually carrying troops began with the homecoming of the 10th BCT, in April 1951, aboard RPS *Cotabato* (ex-USS *LST-75*), one of the five tank landing ships of the Service Squadron of the Philippine Navy. She and its other members—RPS *Pampanga* (ex-USS *LST-842*), RPS *Bulacan* (ex-USS *LST 843*), RPS *Albay* (ex-USS *LST-865*), and RPS *Misamis Oriental* (ex-USS *LST-875*)—transported all but the 14th and 2nd Battalion Combat Teams to and from Korea. The 14th was ferried home, and the 2nd BCT both ways, by U.S. naval vessels.[7]

OTHER ACTIONS IN EARLY DAYS OF THE WAR

At the outset of the Korean War, the ROKN, with 71 vessels of various types, was outnumbered by the 110 vessels of the North Korean navy.

Moreover, the concentration of ROKN units at western and southern ports, enabled the Communists to land ground forces at a few locations along the east coast as far south as Samchok (northeast coast of South Korea just below the 38th parallel) and send the freighter carrying 600 North Korean troops toward the port of Pusan (which is described in Chapter 1).[8]

Unfazed, the ROKN and UN Allies soon drove enemy naval combatants, as well as reinforcement and resupply vessels from South Korean waters. On 2 July, just south of the 38th parallel in the Sea of Japan, light cruiser USS *Juneau*, and British cruiser HMS *Jamaica* and frigate HMS *Black Swan*, sank three of four North Korean PT-boats and both of two motor gunboats in the Battle of Chumonchin Chan. This was the first and last time in the war that enemy naval forces elected to fight the UN navies for control of the sea.[9]

That same day, the ROKN's Naval Base Detachment at Pohang, located on the east coast of South Korea, wiped out a North Korean landing force. The following day, the minesweeper ROKS *Kim Chon* (YMS-513) destroyed three Communist supply vessels near Chulpo on the southwestern coast.[10]

Photo 2-1

Future HMS *BYMS-2258* off Weaver Shipyards, Orange, Texas, 7 June 1944. Courtesy of Dr. Howard C. Williams, Texana and Orange County Historian

The largest production run of any World War II warship was not, as one might imagine, a particular class of destroyer, frigate, or submarine, but instead a class of wooden minesweeper known as the YMS. The class, consisting of 561 scrappy little 136-foot wooden-hulled vessels, was characterized by Arnold Lott in *Most Dangerous Sea* as belligerent-looking yachts wearing grey paint. Of the 561 minesweepers built in American yards, 150 had been transferred directly to the Royal Navy under the provisions of "Lend-Lease." One such was *Kim Chon*, laid down, on 31 October 1942, as *YMS-258*, at Weaver Shipyard, Orange, Texas. She was completed, on 8 February 1944; transferred to Great Britain as HMS *J-1058*; and later reclassified HMS *BYMS-2258*. Their BYMS designation meant, "British" YMS-class ships.[11]

Original armament of the future *BYMS-2258*, before her transfer to the Royal Navy, was: one 3"/50 dual-purpose gun mount, two 20mm mounts, and two depth charge projectors. It is unknown if she retained this armament when she was returned to U.S. custody, on 29 August 1947. She was subsequently struck from the Naval Register, and transferred to South Korea.[12]

ROKN CNO ARRIVES BACK IN SOUTH KOREA

The early successes of the Republic of Korea Navy were particularly notable, considering that chief of Naval Operations Adm. Sohn Won Il was out of the country when war broke out. Sohn was in the United States taking delivery of three ex-USN steel submarine chasers (the former USS *PC-799*, USS *PC-802*, and USS *PC-810*). Following his return to South Korea with the newly acquired *Kum Kang San* (PC-702), *Sam Kak San* (PC-703), and *Chiri San* (PC-704), these warships were quickly pressed into action to oppose the enemy's drive toward Pusan.[13]

Photo 2-2

Kum Kang San off the Mare Island Naval Shipyard, California, 17 June 1950, following her transfer to the South Korean Navy. Naval History and Heritage Command photograph #NH 85483

Less than a week after *Kim Chon* (YMS-513) had destroyed the supply vessels near Chulpo on 22 July, *Kum Kang San* (PC-702) and *Sam Kak San* (PC-703) caught a group of sampans carrying ammunition west of Inchon and sank twelve of them.[14]

Photo 2-3

South Korean submarine chaser *Kum Kang San* (PC-702) at the Mare Island Naval Shipyard, California, 16 June 1950, following transfer to the South Korean Navy. She is flanked by her sister ships *Chiri San* (PC-704) to left and *Sam Kak San* (PC-703). Work is in progress on the 3"/50 dual-purpose gun mounted on *Kum Kang San*'s fo'c'sle. Naval History and Heritage Command photograph #NH 85494

3

Initial United States and Commonwealth Naval Forces

Photo 3-1

Vice Adm. C. Turner Joy, USN (left), commander, U.S. Naval Forces, Far East, and Vice Adm. Arthur D. Struble, USN, commander, Seventh Fleet, aboard the battleship USS *Missouri* (BB-63), off Korea, on 18 October 1950.
Naval History and Heritage Command photograph 80-G-668794

American forces in the Orient in 1950 were a part of the Far East Command under General MacArthur, who was also, as Supreme Commander for the Allied Powers, responsible for the occupation of Japan. Only a little over a third of the U.S. Navy's active strength was in the Pacific, only a fifth of that was in the Far East, and the naval fighting component under Vice Adm. C. Turner Joy, USN, commander, Naval Forces Far East (ComNavFE) was very small.[1]

The ships of Turner's command (designated Task Force 96) in Japanese waters, on 25 June 1950, are identified in the following table. Mine Squadron 3 consisted of six 136-foot, wooden-hulled former YMSs, which the Navy had re-designated as auxiliary minesweepers (AMS), and four 184-foot, steel-hulled *Admirable*-class minesweepers (AM). Three of the latter AMs were in caretaker status (laid up) and the fourth, *Pledge*, in reduced commission.

Naval Force Japan (Task Force 96): Vice Adm. C. Turner Joy, USN

Task Unit 96.5.1 (Flagship Element): Capt. Jesse C. Sowell, USN
light cruiser USS *Juneau* (CL-119)

Task Unit 96.5.2 (Destroyer Element): Capt. Halle C. Allan, Jr., USN
Destroyer Division 91

USS *Mansfield* (DD-728) flagship	USS *Collett* (DD-730)
USS *De Haven* (DD-727)	USS *Lyman K. Swenson* (DD-729)

Task Unit 96.5.3 (British Commonwealth Support Element):
Comdr. Ian Hunter McDonald, RAN
frigate HMAS *Shoalhaven* (F535)

Task Unit 96.5.6 (Submarine Element):
Lt. Comdr. Lloyd V. Young, USN
submarine USS *Remora* (SS-487)

Task Group 96.6 (Minesweeping Group):
Lt. Comdr. Darcy V. Shouldice, USN
Mine Squadron 3

Mine Division 31	Mine Division 32
USS *Redhead* (AMS-34)	USS *Pledge* (AM-277)
USS *Mockingbird* (AMS-27)	USS *Incredible* (AM-249)
USS *Osprey* (AMS-28)	USS *Mainstay* (AM-261)
USS *Partridge* (AMS-31)	USS *Pirate* (AM-275)[3]
USS *Chatterer* (AMS-40)	
USS *Kite* (AMS-22)	

The Australian frigate HMAS *Shoalhaven* was then the single Commonwealth naval vessel under Admiral Joy's command. She began Korean war operations as an escort for American transports running between Sasebo in Japan and Pusan in South Korea. On 7 July, she relieved the American destroyer *De Haven* on the west coast blockade patrol, working with the *Collett* for three days before returning to Sasebo. This was her only patrol of the war. At its conclusion, she resumed her former role of escorting convoys between Sasebo and Pusan. This duty occupied the remainder of *Shoalhaven*'s service in Korean waters, which ended upon her reaching Pusan, on 31 August. There, she relinquished escort of her fourteenth and last convoy and, thence to Kure. A few days later, on 6 September 1950, she sailed from Kure, for Australia.[4]

Fortunately, Task Forces 90 and 96 were not the only naval units in Asiatic waters. Based in the Philippines, under the command of Vice Adm. Arthur D. Struble, was the Seventh Fleet, the embodiment of American naval power in the Western Pacific. Considering the unpredictable responsibilities of Admiral Joy's situation, the transfer of Seventh Fleet forces to his operational control was all that could be done at the time.[5]

However, an important addition soon came in the form of British Commonwealth units commanded by Rear Adm. Sir William G. Andrewes, KBE, CB, DSO RN, who was Flag Officer Second in Command, Far Eastern Station. On 29 June, the British Admiralty placed Royal Navy units in Japanese waters at the disposal of ComNavFE (Admiral Joy). Answering the call, similar action was taken the following day by the Australian government; in Canada, three destroyers were ordered to prepare to sail; and from New Zealand came the promise of the early dispatch of two frigates.[6]

Photo 3-2

Formal portrait of Vice Adm. William G. Andrewes, RN, taken sometime between his 1 December 1950 promotion to Vice Admiral and mid-1952. Naval History and Heritage Command photograph #NH 97139

Commonwealth naval strength in the Far East was significant. Andrewes' Task Group 96.8, on 30 June 1950, included HMS *Triumph*, a 13,000-ton light carrier; two 6-inch gun cruisers, heavily armored *Belfast* (the largest cruiser in the Royal Navy), and *Jamaica*; three destroyers, and four frigates. The destroyer *Bataan*, and previously referenced frigate *Shoalhaven* (now a part of the British Task Unit) were units of the Royal Australian Navy.

British Task Group 96.8	
HMS *Triumph* (R16)	HMAS *Bataan* (I91)
HMS *Belfast* (C35) flagship	HMS *Black Swan* (F116)
HMS *Jamaica* (C44)	HMS *Alacrity* (F57)
HMS *Cossack* (D57)	HMS *Hart* (F58)
HMS *Consort* (D76)	HMAS *Shoalhaven* (F535)[7]

After they had subsequently been placed under U.S. operating command, Admiral Andrewes' ships were allocated between commander, Naval Forces Far East's Support Group (Task Group 96.5) and the Seventh Fleet Striking Force (Task Force 77), which had reached Okinawa, on 30 June. Joined the next day by *Triumph*, *Belfast*, *Cossack*, and *Consort*, Task Force 77 was then poised between Korea and Formosa.[8]

BLOCKADE OF KOREA

Two of the many requirements which faced Admiral Joy in the first hectic days of the war were the provision of escorts for shipping en route to Pusan, and the establishment of a blockade of North Korea. These matters were dealt with by ComNavFE in Operation Order 8-50 promulgated on 3 July, resulting in organizational changes within Task Force 96.[9]

Task Group 96.1 was assigned responsibility for escort of shipping between Japan and Korea, with a commander and units to be assigned when available.[10]

Blockade and inshore work south of latitude 37°N was assigned to the ROK Navy, which soon after became Task Group 96.7, with available assistance from the Far East Air Forces and from any NavFE units in the area. For the coastline north of 37°N, separate East and West Coast Support Groups were established.[11]

East coast blockade and escort duty was assigned to Rear Adm. John M. Higgins' Task Group 96.5, and that in the west to the Commonwealth units of Task Group 96.8. The northern limits of the blockade were set at 41°N on the east coast and at 39°30'N in the west. A precaution implicit in these boundaries, was for all units to keep well clear of Manchurian and Russian waters.[12]

4

Allied Nations' Warships

Ten Allied nations dispatched warships to Korea to operate under the UN flag. A complete list of the naval units of the Republic of Korea (those available at the onset of war, and others acquired during) may be found in Appendix A. Large numbers of United States Navy ships participated in the war, with nearly all operating mostly off the east coast of Korea in the Sea of Japan. Appendix B identifies USN ships that earned battle stars in Korea.

The ships of the other eight Allied navies operated primarily in the Yellow Sea, off the west coast of Korea. Because their service is so closely related to the subject of this book, it is worthwhile to acquaint readers with them at this time. The Allied navies identified in the table contributed sixty-nine ships—sixty-one of them surface combatants.

Royal Navy	Royal Australian Navy	Royal New Zealand Navy	Royal Canadian Navy
5 carriers	1 carrier		
1 depot ship			
1 submarine			
6 cruisers			
7 destroyers	4 destroyers		8 destroyers
14 frigates	4 frigates	6 frigates	

Royal Thai Navy	Royal Netherlands Navy	Colombian Navy	French Navy
	3 destroyers		
4 frigates	1 frigate	3 frigates	1 frigate

These ships, which served as "maids of all duties"—escorting and screening aircraft carriers; conducting naval gunfire bombardment; delivering special forces to hostile beaches; guarding minesweepers during their dangerous duty opening critical ports; and protecting friendly guerilla forces operating from islands behind enemy lines—are identified in the next few pages.

ROYAL NAVY

Light Aircraft Carriers – 4

HMS *Glory* (R62) KOREA 1951-53	HMS *Theseus* (R64) KOREA 1950-51
HMS *Ocean* (R68) KOREA 1952-53	HMS *Triumph* (R16) KOREA 1950

Aircraft Maintenance Carrier – 1

HMS *Unicorn* (I72) KOREA 1950-53

Depot Ship – 1

HMS *Tyne* (A24) KOREA 1953

Submarine – 1

HMS *Telemachus* (S321) KOREA 1953

Light Cruisers – 6

HMS *Belfast* (C35) KOREA 1950-52	HMS *Ceylon* (C30) KOREA 1950-52	HMS *Kenya* (C14) KOREA 1950-51
HMS *Birmingham* (C19) KOREA 1952-53	HMS *Jamaica* (C44) KOREA 1950	HMS *Newcastle* (C76) KOREA 1952-53

Destroyers – 7

HMS *Charity* (D29) KOREA 1950-53	HMS *Concord* (D63) KOREA 1950-53	HMS *Constance* (D71) KOREA 1950-52
HMS *Cockade* (D34) KOREA 1950-53	HMS *Consort* (D76) KOREA 1950-53	HMS *Cossack* (D57) KOREA 1950-53
HMS *Comus* (D43) KOREA 1950-53		

Frigates – 14

HMS *Alacrity* (F60) KOREA 1950-52	HMS *Modeste* (F42) KOREA 1953
HMS *Alert* (F647) KOREA 1951	HMS *Morecambe Bay* (F624) KOREA 1950-53
HMS *Amethyst* (F116) KOREA 1951-52	HMS *Mounts Bay* (F627) KOREA 1950-53
HMS *Black Swan* (F57) KOREA 1950-51	HMS *Opossum* (F33) KOREA 1952-53
HMS *Cardigan Bay* (F630) KOREA 1950-53	HMS *Sparrow* (F71) KOREA 1953
HMS *Crane* (F123) KOREA 1952-53	HMS *St. Brides Bay* (F600) KOREA 1950-53
HMS *Hart* (F58) KOREA 1950-51	HMS *Whitesand Bay* (F633) KOREA 1950-53

Photo 4-1

Light cruiser HMS *Belfast* (C35) coming alongside the light aircraft carrier USS *Bataan* (CVL-29) while operating off the coast of Korea, on 27 May 1952. National Archives photograph #80-G-633883

Photo 4-2

Destroyer HMS *Charity* (D29) off the Korean coast, while covering Operation Fishnet, which was intended to destroy North Korean fishing nets in an effort to reduce Communist forces' food supplies. Photograph is dated 16 September 1952. National Archives photograph #80-G-K-14206

Chapter 4

ROYAL AUSTRALIAN NAVY

Light Aircraft Carrier – 1
HMAS *Sydney* (R17), *Majestic*-class, KOREA 1951-52

Destroyers – 4

HMAS *Anzac* (D59) Battle-class KOREA 1951-53	HMAS *Tobruk* (D37) Battle-class KOREA 1951-53
HMAS *Bataan* (D191) Tribal-class KOREA 1952	HMAS *Warramunga* (D123) Tribal-class KOREA 1950-52

Frigates – 4

HMAS *Condamine* (F698) River-class KOREA 1952-53	HMAS *Murchison* (F442) Bay-class (modified *River*-class) KOREA 1951-52
HMAS *Culgoa* (F408) Bay-class (modified *River*-class) KOREA 1953	HMAS *Shoalhaven* (F535) River-class KOREA 1950

Photo 4-3

Australian destroyer HMAS *Tobruk* (D37) in Korean waters.
Australian War Memorial photograph P05890.070

ROYAL NEW ZEALAND NAVY

Frigates – 6 (*Loch*-class)

HMNZS *Hawea* (F422)	HMNZS *Rotoiti* (F625)
ex-HMS *Loch Eck* (K422)	ex-HMS *Loch Katrine* (K625)
KOREA 1951-53	KOREA 1950-53
HMNZS *Kaniere* (F426)	HMNZS *Taupo* (F421)
ex-HMS *Loch Achray* (K426)	ex-HMS *Loch Shin* (K421)
KOREA 1953	KOREA 1951-52
HMNZS *Pukaki* (F424)	HMNZS *Tutira* (F517)
ex-HMS *Loch Achanalt* (K 424)	ex-HMS *Loch Morlich* (K517)
KOREA 1950	KOREA 1950-51

Photo 4-4

New Zealand frigate HMNZS *Hawea* (F422) during the Korean War. Australian War Memorial photograph P05890.044

ROYAL CANADIAN NAVY

Destroyers – 8	
HMCS *Athabaskan* (DDE219) Tribal-class KOREA 1950-53	HMCS *Huron* (DDE216) Tribal-class KOREA 1951-53
HMCS *Cayuga* (DDE218) Tribal-class KOREA 1950-52	HMCS *Iroquois* (DDE217) Tribal-class KOREA 1952-53
HMCS *Crusader* (DDE228) C-class KOREA 1952-53	HMCS *Nootka* (DDE213) Tribal-class KOREA 1951-52
HMCS *Haida* (DDE215) Tribal-class KOREA 1952-53	HMCS *Sioux* (DDE225) V-class KOREA 1950-52

Photo 4-5

Royal Canadian Navy destroyer HMCS *Athabaskan* during the Korea War. Australian War Memorial photograph P05890.060

ROYAL THAI NAVY

Frigates – 4

HTMS *Bangpakong* ex-HMS *Burnet* (K348), Flower-class corvette (later reclassified frigate)	HTMS *Prasae* ex-HMS *Betony* (K274), Flower-class corvette (later reclassified frigate)
HTMS *Tachin* (PF1) ex-USS *Glendale* (PF-36), *Tacoma*-class frigate	HTMS *Prasae* (PF-2) ex-USS *Gallup* (PF-47), *Tacoma*-class frigate

Photo 4-6

Rescue operations to retrieve survivors of the Thai frigate *Prasae* (ex-HMS *Betony*), which stranded behind enemy lines on the Korean east coast, in a blinding snowstorm, on 7 January 1951. She was destroyed after attempts to pull her off were unsuccessful. A Sikorsky HO3S-1 helicopter of squadron HU-1 is taking off with a load of survivors. Aircrewmen armed submachine guns provide security during the operation. The Thai Navy later acquired the ex-USS *Gallup*, and named her the frigate *Prasae* (PF-2).
Naval History and Heritage Command photograph #NH 97164

Photo 4-7

USS *Glendale* (PF-36) and *Gallup* (PF-47) fly Thai flags, during transfer ceremonies at Yokosuka Naval Base, Japan, 29 October 1951. *Glendale* became HTMS *Tachin* (PF-1) and *Gallup* HTMS *Prasae* (PF-2).
Naval History and Heritage Command photograph #NH 97102

Photo 4-8

HTMS *Bangpakong*, in the Chow Praya River, just below Bangkok, in January 1970.
Naval History and Heritage Command photograph #NH 92665

Other Allied Nations' Naval Forces

ROYAL NETHERLANDS NAVY

Destroyers – 3

HNMS *Evertsen* (D802)	HNMS *Van Galen* (D804)
ex-HMS *Scourge* (G01), S-class destroyer	ex-HMS *Noble* (G84), N-class destroyer
HNMS *Piet Hein* (G94)	
ex-HMS *Serapis* (G94), S-class destroyer	

Frigate – 1

HNMS *Johan Maurits van Nassau* (K251)
ex-HMS *Ribble* (K251), River-class frigate

Photo 4-9

Australian frigate HMAS *Condamine* (K698) at Williamstown, Victoria, circa 1948. She was a *River*-class frigate, like the Dutch HNMS *Johan Maurits van Nassau* (K251). Australian War Memorial photograph 300544

COLOMBIAN NAVY

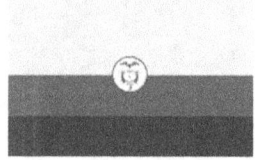

Frigates – 3	
ARC *Almirante Padilla* (F11) ex-USS *Groton* (PF-29)	ARC *Almirante Brion* (F14) ex-USS *Burlington* (PF-51)
ARC *Capitan Tono* (F12) ex-USS *Bisbee* (PF-46)	

Photo 4-10

Colombian frigate ARC *Almirante Brion* (F14) off Coco Solo, Canal Zone, 6 July 1955. Naval History and Heritage Command photograph #NH 81518

Other Allied Nations' Naval Forces

FRENCH NAVY

Frigate – 1
FS *La Grandiere* (F731)
Bougainville-class frigate

Photo 4-11

French frigate FS *La Grandiere* (F731), location and date unknown.
Courtesy of University of Otago Library, New Zealand, photograph S06-25j

5

Naval Blockade and Escort Duties

> *Enemy action was not the only obstacle facing the blockading ships; there were also the problems created by geography, hydrography and climate. The western coastline, for instance, is ragged and heavily indented, and the water is extremely shallow and dotted with islands, low-water mud flats, rocks and shoals. High, strong tides, of over thirty feet in some places, scour the muddy bottom, and channels are formed, obliterated and reformed with remarkable frequency.*
>
> —Thor Thorgrimsson and E. C. Russell in *Canadian Naval Operations in Korean Waters 1950-1955*.[1]

On 4 July 1950, President Harry S. Truman declared a blockade of the entire Korean coastline, and it was up to UN naval forces to make it effective. Any threat posed by North Korea's small, ineffective "gunboat navy," and obsolete planes that made up its air force, was quickly eliminated. There remained, however, the dangers sea mines and shore-battery fire posed to Allied ships. As expressed in the quoted material, geography presented challenges to the safe navigation of ships, particularly in the Yellow Sea off the west coast of Korea. In addition to prevalent shoal water along the coastline, and in the vicinity of islands, the few harbors on the west coast worthy of that name had to be continually dredged to prevent silting.[2]

In contrast these conditions, which hindered efforts by the west coast blockade force to carry out its duties, were quite favorable for enemy clandestine use of small craft to land agents ashore, and move supplies and men by sea to and from the mainland and offshore islands. The shallows made it difficult for even small Allied ships to approach close enough to the shore to provide effective gunfire support for UN and guerilla forces, and to attack the enemy's supply lines. Finally, conditions on the west coast made it easy for the enemy to lay mines. Fortunately, while the extreme range of the tides hid moored mines at high tide, they were fairly easy to locate at low tide as a result of the greatly decreased water column. However, unmoored floating mines presented a perennial problem.[3]

ADDITIONAL ALLIED SHIPS ARRIVE IN THEATER

As destroyers and frigates from Canada, New Zealand, France, and the Netherlands joined American, ROK, British, and Australian warships already in theater, Allied naval forces grew steadily stronger. The following table shows all Allied ship arrivals and departures (excluding USN and ROK) from June through December 1950:

HMS: His/Her Majesty's Ship
HMAS: His/Her Majesty's Australian Ship
HMCS: His/Her Majesty's Canadian Ship
HMNZS: His/Her Majesty's New Zealand Ship
HNMS: His/Her Netherlands Majesty's Ship

FS: French Ship
HTMS: His Thai Majesty's Ship

1950

Month	Arrived	Departed
June	HMS *Alacrity* (F60)	
	HMS *Belfast* (C35)	
	HMS *Black Swan* (F57)	
	HMS *Consort* (D76)	
	HMS *Cossack* (D57)	
	HMS *Hart* (F58)	
	HMS *Jamaica* (C44)	
	10 - HMAS *Bataan* (D191)	
	27 - HMAS *Shoalhaven* (F535)	
July	HMS *Charity* (D29)	
	HMS *Cockade* (D34)	
	HMS *Comus* (D43)	
	HMS *Kenya* (C14)	
	3 - HMNZS *Pukaki* (F424)	
	3 - HMNZS *Tutira* (F517)	
	16 - HNMS *Evertsen* (D802)	
	29 - FS *La Grandiere* (F731)	
	30 - HMCS *Athabaskan* (DDE 219)	
	30 - HMCS *Cayuga* (DDE218)	
	30 - HMCS *Sioux* (DDE225)	
August	HMS *Alert* (F647)	HMS *Alacrity* (F60)
	HMS *Ceylon* (C30)	HMS *Belfast* (C35)
	HMS *Mounts Bay* (F627)	HMS *Black Swan* (F57)
	HMS *Whitesand Bay* (F633)	HMS *Comus* (D43)
	14 - HMAS *Warramunga* (D123)	HMS *Hart* (F58)
September	HMS *Concord* (D63)	22 - HMAS *Shoalhaven* (F535)
October	HMS *Constant* (D71)	HMS *Alert* (F647)
	HMS *Morecambe Bay* (F624)	HMS *Jamaica* (C44)
	7 - HMNZS *Rotoiti* (F625)	
November	HMS *Cardigan Bay* (F630)	HMS *Cockade* (D34)
	7 - HTMS *Bangpakong*	HMS *Mounts Bay* (F627)
	7 - HTMS *Prasae*	25 - FS *La Grandiere* (F731)
December	HMS *Mounts Bay* (F627)	HMS *Whitesand Bay* (F633)
	HMS *St. Brides Bay* (F600)	3 - HMNZS *Tutira* (F517)

The dates in the table reflect when ships arrived and departed the theater in 1950, or were relieved of their duties by the ship that replaced them. Entries for ships of the Royal Navy reflect the month they began a patrol in Korean waters, and when it completed. The intervening periods do not include time spent in port, for ship provisioning, maintenance, or crew rest. (Appendix E offers ship deployment summary information for 1950-1955.)

EVACUATION OF SOUTH KOREAN CAPITAL CITY

Involvement of U.S. naval forces in the Korean War began in late June. With the rapid North Korean advance southward from the 38th parallel, the U.S. ambassador to Korea, John J. Muccio, requested an immediate evacuation of U.S. nationals from Seoul. Admiral Joy responded quickly. On 26 June 1950, the destroyers *Mansfield* (DD-728) and *De Haven* (DD-727) arrived at the nearby port of Inchon, to which Americans from Seoul had been transported that morning. By late afternoon, 700 U.S. and friendly foreign nationals had crowded aboard the Norwegian-flag freighter *Reinholt* in the harbor, and she weighed anchor, under escort by the destroyers.[4]

Air cover for the ensuing seaborne evacuation to Yokosuka, Japan, was provided by U.S. Air Force fighters operating from airbases that would shortly be overrun by North Korean troops.[5]

Those who could not find places aboard *Reinholt* were subsequently evacuated by airlift from Kimpo Airfield, near Seoul, or from Suwon, farther south. About 750 U.S. citizens and friendly foreign nationals were evacuated by air, on 27 June, and another 850 on 28 June. Other U.S. nationals from Taejon, Taegu, and Pusan made their escape by ship from that port. By 29 June, the evacuation had been completed without a single casualty, just in time; Seoul fell, on 27 June, to the North Korean Army.[6]

CARRIER TASK FORCE 77 SAILS FROM OKINAWA

On the evening of 1 July, Task Force 77—the carriers USS *Valley Forge* (CV-45) and HMS *Triumph* (R16), escorted by two cruisers, and ten destroyers—departed Buckner Bay, Okinawa. After proceeding northwest and north to a designated launch position in the Yellow Sea, the task force commenced air strikes into North Korea, on 3 July.[7]

Off the east coast, HMS *Black Swan* gained some notoriety that same day (evening of 3 July), by being the first Allied ship to come under enemy air attack. Two enemy fighters, thought to have been Soviet-built Sturmovik aircraft, came in on the British frigate from over the land and out of the haze. They inflicted minor structural damage, and escaped without being hit by gunfire. Fortunate in their evasive action, the pilots of these aircraft were doubly blessed in their assignment. While they were away from their base; their colleagues at Pyongyang received a thorough working over by the aircraft of Task Force 77. *Black Swan*'s experience remained unique until 22 August, when the British destroyer HMS *Comus* underwent an attack from the air.[8]

Photo 5-1

HMS *Black Swan* at Kure, Japan, circa 1950-55.
Australian War Memorial photograph 146376

ANDREWES' WEST COAST GROUP GROWS IN SIZE
In mid-July, a ComNavFE reorganization assigned all blockade ships to Task Group 96.5, commanded by Rear Adm. Charles C. Hartman. Designated the East Coast Support Group, it was split into two alternating elements, CTE 96.51 and CTE 96.52. Rear Admiral Andrewes' West Coast Support Group was re-designated CTE 96.53, and an additional unit, the Escort Element, CTE 96.50, was formed.

Commander, Naval Forces Far East Vice Adm. C. Turner Joy, USN	
Task Group 96.5 Rear Adm. Charles C. Hartman, USN	
Task Element 96.50 Capt. Alan D. H. Jay, RN, in HMS *Black Swan* (Escort Element of DDs and Frigates)	Task Element 96.51 Rear Adm. Charles C. Hartman, USN East Coast Support Group No. 1
Task Element 96.53 Rear Adm. William Andrewes, RN West Coast Support Group[9]	Task Element 96.52 Rear Adm. John M. Higgins, USN East Coast Support Group No. 2

The Escort Element (TE 96.50), under Capt. Alan D. H. Jay, RN, consisted of the British frigates HMS *Black Swan* and HMS *Hart*, and the Australian frigate HMAS *Shoalhaven*.[10]

A principal change in the new command organization was giving the British responsibility for the west coast of Korea while the United States Navy looked after the east coast. According to Admiral Joy, the reasons for this arrangement were:

> ... purely tactical in nature. For one reason, the east coast with its longer coastline and more numerous accessible targets required more ships for blockade, as well as bombardment and interdiction missions, than the British could muster. Furthermore, since our fast carriers would be operating most of the time in the Sea of Japan it was thought best from the standpoint of coordination to have U.S. ships rather than British operating in the same area as the carriers.[11]

By mid-July, Andrewes had under his command for Yellow Sea operations the British light cruisers HMS *Jamaica*, *Kenya*, and *Belfast*, and destroyers *Cossack*, *Cockade*, *Charity*, the Australian destroyer HMAS *Bataan*, and the Netherlands destroyer HNMS *Evertsen*. This force was bolstered further in late July with the arrival of three Canadian destroyers, HMCS *Cayuga*, *Athabaskan*, and *Sioux*. On 1 August, the French frigate FS *La Grandiere* and two New Zealand frigates, HMNZS *Tutira* and *Pukaki*, joined the UN force.[12]

The additional ships were most welcome, and necessary. ComNavFE had directed, on 10 July, that the naval blockade be extended to the North Korean ports of Wonsan on the east coast, and Chinnampo on the west coast.[13]

INAUGURATION OF THE WEST COAST BLOCKADE

HMS *Cossack*   **HMS *Belfast***   **HMS *Consort***

On 8 July in an operation order, Admiral Andrewes outlined his intentions for operations—mainly focused on the west coast blockade.

Chapter 5

- Enforcement of blockade of the coast occupied by North Koreans
- Prevention of infiltration by sea on coasts held by South Koreans
- Provision of naval support as required against North Korean maritime forces or land targets[14]

To carry out these duties, Admiral Andrewes split his then limited blockade naval forces into three task units:

TU	Commander	Ships
96.8.1	Rear Adm. Andrewes	HMS *Belfast*, HMS *Cossack*, HMS *Consort*
96.8.2	Capt. Jocelyn Salter	HMS *Jamaica*, HMS *Black Swan*, HMS *Alacrity*
96.8.4	Capt. Patrick Brock	HMS *Kenya*, HMS *Cockade*[15]

Operating separately, each of these units carried out patrols in the blockade area in rotation, as ordered. As Andrewes sailed from Sasebo, on 9 July, for the Korean west coast, embarked in the light cruiser *Belfast*, with the destroyers *Cossack* and *Consort* in company on the first patrol, the blockade was inaugurated.[16]

Map 5-1

Korea and relatively nearby ports of Sasebo and Kure, Japan [Sasebo and Kure are located in southwest Japan; the former in the Nagasaki Prefecture, and the latter (not shown on map) on the Inland Sea in Hiroshima Prefecture]

HMS *COSSACK* BRACKETED BY ENEMY GUNFIRE

Photo 5-2

British destroyer HMS *Cossack*, location and date unknown.
Imperial War Museums photograph FL 7087

After a stop at Pusan to pick up a South Korean liaison team, whose personnel were distributed among the three British warships, 96.8.1 proceeded to the west coast of Korea. Apart from encountering bad weather, the passage north of Quelpart Island (Cheju-do on Map 5-1), was uneventful.[17]

On 10 July, the unit made a sweep off Inchon, sufficiently close in to sight the main entrance channel into the port and the Flying Fish Channel, and to explore the islands off the entrance. Finding only a few junks engaged in fishing, the ships proceeded seaward, then turned northwestward to search for enemy vessels that might be in passage between Chinnampo and Inchon. Nothing was sighted, and it appeared that the sea was clear of anything larger than small fishing vessels.[18]

While patrolling northward along the coast between Inchon and Chinnampo, the ships proceeded to the entrance of the latter port. The southern limit of the entrance was well marked, but to the north were endless shoals and shallows. The port lay over 20 miles up a narrow and torturous approach channel. In early morning, on 12 July, *Cossack* was detached at 0600 to search shoreward of the Taechong Islands (see Map 5-3), and was fired on by shore batteries at Paenyong-do.[19]

A short but hot engagement followed, in which *Cossack* (under the command of Capt. Varyl C. Begg, DSC RN), after firing about 140 rounds at ranges of 5,000 to 8,000 yards, silenced two shore guns. Ten or more enemy rounds fell within 200 yards of the destroyer, but she suffered no harm or casualties.[20]

Map 5-2

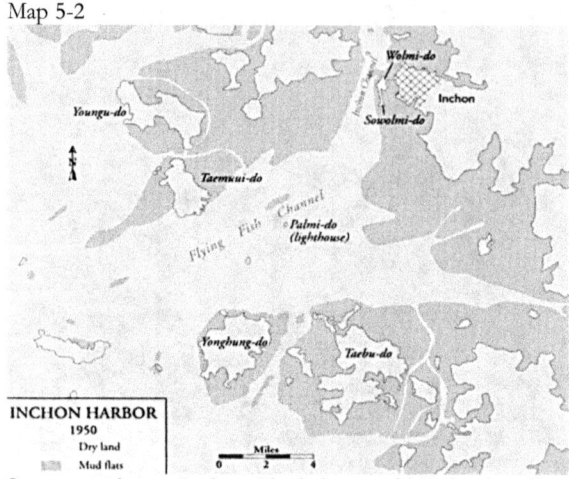

Sea approaches to Inchon (shaded areas along the shoreline, around islands, and in other areas, represent mud flats)

Map 5-3

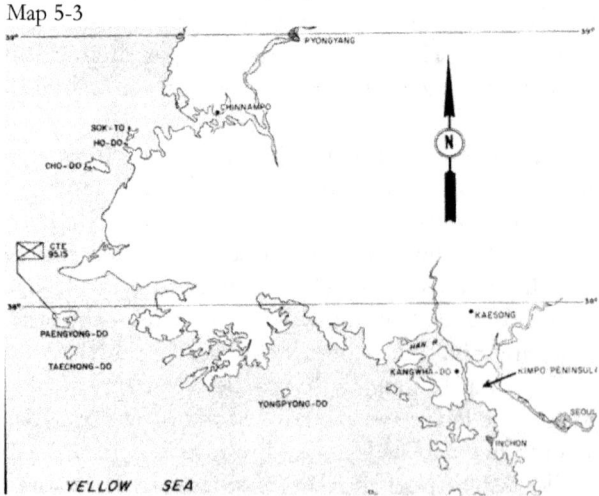

West coast of Korea between Pyongyang and Inchon in summer 1952 (CTE 95.15 refers to the West Coast Defense Element then present on Paengyong-do, near Taechong-do) Pat Meid and James M. Yingling, *U.S. Marine Operations in Korea 1950-1953 Vol. V. Operations in West Korea*

Prior to this time, naval commanders had been unaware of enemy field artillery on these islands. Although there was no further shelling of Allied forces from Communist-held islands following this incident, the threat of such was always a concern. On 13 July, the ships set a course southward, and Capt. Jocelyn S. C. Salter's group took over the patrol that afternoon.[21]

ESCORT DUTY BETWEEN SASEBO AND PUSAN

Photo 5-3

At Sasebo, Japan, officers and men of Destroyer Division 91 assembled topside aboard their ships in December 1951, to receive the Navy Unit Commendation. The famed 'Sitting Duck' destroyers are berthed in their numerical order: USS *De Haven* (DD-727), *Mansfield* (DD-728), *Lyman K. Swenson* (DD-729), and *Collett* (DD-730).
Naval History and Heritage Command photograph #NH 97090

Following her arrival in theater, on 27 June 1950, the Australian frigate HMAS *Shoalhaven* (F535) had been assigned escort for American transports running between Sasebo in Japan and Pusan, South Korea. On 7 July, she relieved the destroyer USS *De Haven* (DD-727) on the west coast blockade patrol, and worked with USS *Collett* (DD-730) for three days before returning to Sasebo. (Prior to Andrewes' task force undertaking blockade duties, US destroyers had carried out this work.)

This interlude was *Shoalhaven*'s only patrol of the Korean War, before resuming her previous role of escorting convoys between Sasebo and Pusan. Escort duty ended for her, on 31 August, upon reaching Pusan with her fourteenth and last convoy. Following a visit to Kure, she departed Japanese waters, on 6 September 1950, to return to Australia.[22]

De Haven and *Collett*—along with *Mansfield* and *Swenson*—were members of Destroyer Division 91. Based in San Diego, the division had left there, in late April 1950, on a deployment to the Far East. The ships were in Yokosuka when war broke out and they, like HMAS *Shoalhaven*, were immediately pressed into escort duties. Former *Collet* crewmember Ed Shumer explained the necessity to surge troops and materiel by sea to Pusan, and for subsequent offensive operations against the enemy:

> It didn't take long before the North Koreans had taken most of the South Korean territory. At that time there were not many U.S. warships in the Far East. But then we got the word that ships were on their way. Our main task was getting supplies into Pusan. Any ship, boat, or barge that could haul supplies was used; our job was to protect the supplies. Soon, the USS *Boxer* (CV-21), USS *Toledo* (CA-133), and others were there to help the Navy expand their operations.
>
> With the help of the *Toledo*, we targeted areas along the coastline in late July and early August 1950. Bridges, railroads, and supply yards were often attacked to disrupt supplies going south. The towns of Pohang, Yongdok, Sanechok, Songin, and Iwon are not as well-known as Inchon or Wonsan, but nevertheless they were in the war zone and were attacked often. After a month or so in the war zone we would go back to either Sasebo or Yokosuka, Japan, for a few days to relax. Liberty was always port and starboard [half the crew aboard each day, and the other half allowed liberty ashore].[23]

REPUBLIC OF KOREA NAVAL OPERATIONS

While units of the U.S. Seventh Fleet, and Commonwealth ships under British command, were engaged in offensive operations, so too were units of the ROK Navy. As discussed in Chapter 2, on 22 July, the minesweeper *Kim Chon* (YMS-513) sank three North Korean supply craft near Chulpo in southwestern South Korea. A few days later, on 27 July, the newly acquired sub-chasers *Kum Kang San* (PC-702) and *Sam Kak San* (PC-703) ventured even farther north up the west coast of Communist-controlled South Korea and bombarded Inchon Harbor. The two ships then caught numerous North Korean sampans loaded with ammunition, west of Inchon, and sank twelve of them. In the first week of August, *Kyong Chu* (YMS-502) and other ROKN units destroyed thirteen Communist logistics craft off the west coast of South Korea.[24]

Between August 13 and 21, the ROKN engaged enemy seaborne supply attempts five times and, in one instance, *Kwang Chu* (YMS-503) sank fifteen craft and captured thirty that were trying to supply North Korean Army forces attacking the Pusan Perimeter.[25]

ADDITIONAL ALLIED WARSHIPS JOIN EFFORTS

On 24 July, Admiral Joy, dissatisfied with the ad hoc escort arrangement then existing, established a dedicated Escort Element under Royal Navy Capt. Alan D. H. Jay. As other Allied destroyers and frigates arrived in theater, most were assigned initially to escort duties, bolstering the efforts of HMS *Black Swan*, HMS *Hart*, and HMAS *Shoalhaven*.[26]

The French frigate FS *La Grandiere* (F731) reached Korean waters, on 29 July, and took up escort duties between Sasebo and Pusan. Twelve successive ship convoys (transporting troops and various cargoes) followed, through 3 September. *La Grandiere* was then assigned to a group of frigates and destroyers from various countries. Capt. John H. Unwin, RN, of the frigate HMS *Mounts Bay* (F637), was in command of this group. Later *La Grandiere*'s duties included, in addition to escort work, close inshore support bombardments with HMCS *Athabaskan*, HMCS *Cayuga*, and HMAS *Bataan*.[27]

Photo 5-4

Drawing of Mokpo in the far southwest of the Korean peninsula during an attack by planes from Task Force 77 by Herbert C. Hahn, circa 1951. Naval History and Heritage Command Accession #88-191-BE

Regarding the later duty, on 1 August, HMAS *Bataan*, HMS *Belfast*, and HMS *Charity* had bombarded shore batteries at Haeju Man on the west coast of South Korea, involving a brief exchange with enemy guns. In other action, RN destroyers *Cockade* and *Cossack* bombarded Mokpo Harbor on the west coast, in response to intelligence that many North Korean supply craft were there. A U.S. Navy PV2 Neptune patrol aircraft, assigned to spot the fall of gun rounds, determined that most of the vessels were gone, but docks and railroad sidings were taken under fire with good result. Shore bombardment missions were challenging because extreme tidal variance, sometimes as much as 30 feet, necessitated lengthy approaches in constrained periods during a high tide, and in poorly charted waters.[28]

CANADIAN DESTROYERS ARRIVE IN THEATER

> *You are to sail* Cayuga, Sioux *and* Athabaskan *from Esquimalt at 16 knots to Pearl Harbor p.m. Wednesday 5 July, 1950...*
>
> —Excerpt from a message received by Flag Officer Pacific Coast, on 30 June 1950, from CANAVHED (Naval Headquarters, Ottawa).[29]

On both sides of the Pacific the invasion of South Korea had been followed by a period of furious activity, as Allied nations committed military forces, or other support to the war zone. The government of Canada chose the Royal Canadian Navy (RCN) as its nation's initial military contribution because, of the three armed services, only they were in a position to provide active forces for immediate use. Following this decision, the RCN cancelled a scheduled European cruise for its Pacific Destroyer Division ships—HMCS *Cayuga*, HMCS *Athabaskan*, and HMCS *Sioux*—and, instead, ordered them to the Far East. In execution of their deployment orders, the three destroyers stood out of Esquimalt, in mid-afternoon on 5 July, bound for Pearl Harbor.[30]

Within two hours of their arrival in Hawaii, on 12 July, the ships received orders placing them under "the operational control of General MacArthur as Commander United Nations Forces Korea for operations in relation to the invasion of South Korea only." Leaving Pearl Harbor, the destroyers stopped at Kwajalein and Guam en route to Japan. Upon their arrival at Sasebo, on 30 July, the ships were allotted to the British task elements, as was the convention. HMCS *Cayuga* was assigned to the West Coast Support Group (TE 96.53) and HMCS *Athabaskan* and HMCS *Sioux*, to the Escort Element (TE 96.50).[31]

With these assignments, the three ships of the Canadian Destroyer Division Pacific virtually ceased to exist as a single unit for operational purposes in Korea. For the duration of the war, only occasionally would the three Canadian destroyers, whose commanding officers are identified below, serve together under Canadian command.

- Capt. Jeffry V. Brock, DSC RCN, commander, Canadian Destroyers Pacific, and commanding officer, HMCS *Cayuga*
- Comdr. Paul Dalrymple Taylor, RCN, commanding officer, HMCS *Sioux*
- Comdr. Robert Phillip Welland, DSC RCN, commanding officer, HMCS *Athabaskan*[32]

HMCS *Athabaskan* was the first to go to work. In mid-afternoon on 31 July, she cleared Sasebo Harbor to escort the troop ship USNS *General C. G. Morton* (T-AP-138) to Pusan. This was the first of four convoy missions by *Athabaskan* before her transfer to the west coast group, on 11 August. *Sioux* carried out only one seven-hour patrol in the approaches to Sasebo during the time she was attached to the escort element. On 12 August, she was transferred and joined *Athabaskan* in TU 96.53.3 for service on the west coast.[33]

Cayuga sailed from Sasebo, on the evening of 3 August, as escort to the Royal Fleet Auxiliary (RFA) *Brown Ranger*, sent out to refuel the ships of TU 96.53.3.9. After returning to Sasebo, early on 8 August, from what proved to be an uneventful operation as far as enemy opposition, *Cayuga* carried out five more routine convoy missions to Pusan between the 9th and the 24th, before joining the blockading forces on the west coast.[34]

FIRST RCN OFFENSIVE ACTIONS AGAINST ENEMY

On 15 August, returning from one of her escort assignments to Pusan, *Cayuga* joined the British frigate HMS *Mounts Bay* south of Yosu, a port on the southern Korean peninsula to the west of Pusan for a bombardment operation. The military position of the UN forces within the Pusan Perimeter was still very precarious. Athough the main impetus of the North Korean attack had been slowed, the enemy was still pushing upon the perimeter at most points. One of the more dangerous thrusts was coming from the Yosu and nearby Sunchon areas upon the weakly defended UN left flank. Enemy forces inland were out of range of naval gunfire, but Yosu, recently seized by the North Koreans, was on the coast and therefore vulnerable.[35]

Photo 5-5

Canadian destroyer HMCS *Cayuga* off Korea, 3 March 1954.
National Archives photograph #80-G-642748

Photo 5-6

British frigate HMS *Mounts Bay*, probably at Kure, Japan, circa 1951-1952.
Australian War Memorial photograph P05890.039

There was very little likelihood that the Communists could, in the face of the naval blockade, make any use of Yosu as a port, but there were numerous warehouses and other installations that might be of value to their war effort. Accordingly, HMS *Mounts Bay* and HMCS *Cayuga* had been detailed for the task of destroying them.[36]

The two ships closed to within four miles of the port, anchored, and prepared for action. At 1742, *Mounts Bay*, assisted by an aircraft spotting for her, opened fire and began a methodical bombardment of the harbor area with her 4-inch guns. About half an hour later *Cayuga* joining fire began placing 94 rounds of 4-inch high explosive upon the installations. After almost two hours of successful shelling, the two ships broke off the action and returned to Sasebo to continue their normal escort duties.[37]

When *Cayuga* left the escort element to join the blockade and bombardment force, on 24 August, the other Canadian destroyers had already been serving with it for almost two weeks. *Athabaskan* had been particularly busy working among west coast islands, firing on enemy batteries, observation posts, troop concentrations, and other targets. She also had supported landings of ROK troops on enemy-dominated islands. *Sioux* had carried out a minor bombardment of the town of Popsong'po (Popusompu), on 20 August.[38]

NEW ZEALANDER AND DUTCH WARSHIPS ARRIVE

On 29 June, the government of New Zealand offered two frigates for service in Korea, HMNZS *Tutira* and *Pukaki*. They sailed from Auckland, on 3 July. Upon arrival at Sasebo, on 2 August, they joined other Commonwealth ships of the Escort Element, and took up their duties conveying supply ships between Japan and Pusan.[39]

The Netherlands government offered the destroyer HNMS *Evertsen*, on 3 July, as part of the UN's maritime forces. She left Surabaya, Indonesian (at the time the Netherlands East Indies), for Japan, on 7 July. *Evertsen* arrived at Sasebo, on 16 July, and departed on her first patrol four days later. She was assigned to Task Group 96.5 (Korea-Japan Support Group) which, on 12 September 1950, was re-designated as Task Force 95 (United Nations Blockade and Escort Force).[40]

Task Force 95 units conducted patrols along the entire Korean coast, searching for mines and enemy ship movements. Destroyers and frigates operating along the west coast of Korea, were also tasked with escorting a US or British aircraft carrier on a regular basis. These duties included functioning as a "Bird Dog" or "Plane Guard." Bird Dog duties involved positioning the assigned ship between the carrier and the coastline in order to retrieve any pilots downed at sea out of the water as quickly as possible. Similarly, the role of a Plane Guard ship involved stationing behind a carrier, so as to be available to rescue any pilots crashing into the sea on take-off or landing.[41]

ACTION IN THE YELLOW SEA

In the Yellow Sea, Admiral Andrewes' element (then divided into three rotating sections composed of a cruiser and two or more destroyers each) carried out bombardment and blockade duties. However, as the land warfare moved farther inshore beyond the range of naval guns, no fire support was possible. This elimination of duty allowed greater emphasis on the interdiction of any enemy movements by sea around headlands. Three barrier stations were established off the western headlands between latitude 38°8' and 36°45', on 5 August, and kept activated as availability of ships permitted. Concurrently, the ROK Navy focused on interdiction efforts among the numerous islands and shoal water that fringed the coast.[42]

The principal excitement in August, which came in late month, was the appearance of two enemy aircraft. The first surprised and attacked the destroyer HMS *Comus*, on the 22nd, damaging the ship and killing one of her crew, Leading Stoker Mechanic James W. Addison. The second aircraft attacked an ROK vessel the next day. The other action involved ROKN units sinking numerous enemy small craft.[43]

Facing increased Allied naval presence in west coast waters, the enemy, for the present, abandoned efforts to bring supplies by larger vessels down from the north. Instead, in the south and southwest, vigorous attempts were made to deliver materiel and troops by small boat to the enemy's southern flank. These efforts led to an upsurge in the inshore operations of the ROK Navy. On 20 and 21 August, off Chindo Island on the southwest corner of the Korean peninsula, *Kwang Chu* (YMS-503) found considerable activity. She engaged three enemy motor craft of between 30 and 100 tons; capturing one, sinking one, and damaging the third. Well northward, twenty miles off Inchon, *Bak Du San* (PC-701) sank a large sailing vessel.[44]

In a small estuary east of Chindo, *Guwolsan* (YMS-512) sank one 100-ton motor craft and another of 70 tons, drowning full loads of enemy troops on both. Off Namhae Island on the south coast, *Gaeseong* (YMS-504) damaged 14 of 15 small sailboats she encountered. The most notable work was done by *Kil Chu* (YMS-514) when, in three separate engagements in less than three hours, she sank three enemy vessels and damaged eight. On 31 August, *Kum Kang San* (PC-702) sank two large motor craft and damaged another near Chindo.[45]

6

Inchon Landings (Operation Chromite)

The total strength of Joint Task Force 7 amounted to some 230 ships of all shapes and sizes, from APDs [high-speed transports] of 2,100 tons full load displacement to transports of ten times that size. Except for a few gunnery ships held back to support the flanks of the perimeter, it included all combatant units available in the Far East. Fifty-two ships were assigned to the Fast Carrier, Patrol and Reconnaissance, and Logistic Task Forces; the remainder went to make up the Attack Force, Task Force 90, under Admiral Doyle. Of these, more than 120 were required to lift X Corps, while the rest were involved in gunfire and air support, screening, minesweeping, and miscellaneous other duties.

—James A. Field Jr., *History of United States Naval Operations: Korea.*[1]

The Navy and the Marines have never shone more brightly than this morning.

—Message sent by Gen. Douglas MacArthur, on 15 September 1950, following the initial amphibious landings at Inchon, Korea.[2]

As the North Korean Army drove UN forces southward, MacArthur noted its over-extended supply lines, most of which passed through Seoul. If the city of Inchon, only fifteen miles west of the capital, could be seized by sea assault, the enemy's supply lines would be quickly severed, which would shorten the war, save countless casualties, and possibly eliminate the necessity for a winter campaign. The basis for the general's strategy was his belief in the importance of striking decisively and unexpectedly well behind enemy lines. In addition to preferring Inchon for its location, MacArthur believed that his forces would prevail at that improbable site because the North Koreans would consider a landing there impossible, even insane, and be taken by surprise.[3]

The reason for his belief were the myriad hazards to navigation imposed in the area by geography. The tides of Inchon (33 feet at their maximum) are among the highest in the world and reach their peak in approximately six hours, producing a five-knot current. Vast mudflats near Inchon extend at low water some 6,000 yards seaward. The approach channel to the port city, Flying Fish Channel, is narrow, tortuous, and difficult even for a daylight passage.[4]

Photo 6-1

Gen. Douglas MacArthur on board the command ship USS *Mount McKinley* (AGC-7) during the Inchon landings, 15 September 1950. The others present are (from left to right): Rear Adm. James H. Doyle, USN, commander, Task Force 90; Brig. Gen. Edwin K. Wright, USA, MacArthur's Operations Officer; and Maj. Gen. Edward M. Almond, USA, commander, Tenth Corps.
Naval History and Heritage Command photograph SC 348448

The absence of aids to navigation and the presence of enemy gunfire and mines would make the passage of an invasion fleet through such a channel extremely dangerous. The channel was so narrow that if a ship foundered in the final approach to Inchon, the vessels astern of it would be blocked by those ahead, and become trapped, particularly at low tide. The tides also controlled the invasion date. The tank landing ships would require at least 29 feet of water beneath their keels to reach the selected landing beaches—conditions that existed only a few days each month. Possible dates for such a landing, in the fall of 1950, were limited to 15 September, 11 October, or 3 November, give or take a day or two.[5]

MINE THREAT UNDERESTIMATED, LUCK PREVAILS

As early as 10 July and unknown to the UN commanders, shipments of Soviet sea mines were rolling southward down the Korean east coast railway from the Vladivostok region. Their delivery to North Korea was the first step of a minelaying campaign designed to threaten UN naval forces and make Korean coastal waters untenable to navigation. One week later, Soviet personnel arriving at Wonsan and Chinnampo, were instructing their North Korean counterparts in the use of these mines.[6]

Some 4,000 mines were quickly passed through Wonsan (a major port and transportation hub on the northeast coast of Korea), and by 1 August, mining was begun at that port and at Chinnampo on the west coast. Russian naval officers travelled as far south as Inchon, to oversee preparations for commencement of minelaying; shipments of mines were trucked south from Chinnampo to Haeju while consignments reached Inchon, Kunsan, and Mokpo by train.[7]

Lacking a navy and any physical presence of one, the enemy's effort to contest UN control of the sea went undetected. In mid-August, patrol planes reported barges and patrol craft present at Wonsan and Chinnampo. In retrospect, they were believed to have been engaged in minelaying, but the intelligence was not interpreted as such at the time. Operational planners, while crediting the enemy with limited mining capabilities at Inchon, stated that available information indicated no minefields in these other areas.[8]

On the morning of 10 September, five days before the planned landings at Inchon, the sub-chaser *Sam Kak San* (PC-703) was a little north of Inchon, when her commanding officer, Comdr. Lee Hung So, ROKN, sighted a small boat laying mines. Reacting quickly, he ordered his gun crew to open fire, destroying the enemy vessel with one well-placed round.[9]

This event added to developing concerns regarding the possibility of encountering enemy-emplaced "shipkillers" at Inchon as, less than a week earlier, on 4 September, the destroyer *McKean* (DD-784) spotted mines off Chinnampo. The British cruiser HMS *Jamaica* (C44) and destroyer HMS *Charity* (D29) subsequently sank "floaters" in the same area three days later.[10]

MINES FOUND/DESTROYED DURING PRE-INVASION SHORE BOMBARDMENT

In preparation for the main landings, on 15 September, U.S. Marines were to conduct an initial landing on Wolmi-do Island, in the approaches to Inchon, because of its commanding presence in relation to the shoreline. This was to be followed by an amphibious assault on Red, Yellow, and Blue beaches at Inchon. The U.S. Marine Corps 1st and 5th Divisions would first establish a beachhead and then seize Kimpo Airfield and the Han-Gang River line on the outskirts of Seoul before taking the capital city. The U.S. Army 7th Infantry Division would arrive by convoy after D-Day (it landed on 17 September) and then carry out combat operations together with the Marines as directed by Maj. Gen. Edward M. Almond, U.S. Army commander, as part of Tenth Corps. Established by MacArthur for the invasion, the corps, a joint Marine Corps-Army entrerprise, was comprised of the 1st Marine Division and 7th Infantry Divisions.[11]

The pre-invasion bombardment of Wolmi-do Island commenced early on the morning of 13 September in clear weather and calm seas by Gunfire Support Group 90.6, comprised of four cruisers and six destroyers, as they started up Flying Fish Channel. A few miles south of Inchon, the cruisers anchored in their bombardment stations, while the destroyers continued northward.[12]

These cruisers and destroyers are identified in the following table. (A list of all the Allied cruisers, destroyers, and frigates that participated in the Inchon landings may be found in Appendix F.)

Task Group 90.6 (Gunfire Support Group): Rear Adm. John M. Higgins, USN	
90.6.1 Fire Support Unit 1	Rear Adm. John M. Higgins, USN
USS *Toledo* (CA-133)	Capt. Richard F. Stout, USN
USS *Rochester* (CA-124)	Capt. Edward L. Woodyard, USN
HMS *Kenya* (C14)	Capt. Patrick W. Brock, RN
HMS *Jamaica* (C44)	Capt. Jocelyn S. C. Salter, DSO, OBE RN
90.6.2 Fire Support Unit 2	Capt. Halle C. Allan Jr., USN
USS *Mansfield* (DD-728)	Comdr. Edwin H. Headland, USN

USS *De Haven* (DD-727)   Comdr. Oscar B. Lungren, USN
USS *Lyman K. Swenson* (DD-729) Comdr. Robert A. Schelling, USN
                                 Comdr. Robert H. Close, USN

90.6.3 Fire Support Unit 3

USS *Collett* (DD-730)   Comdr. Robert H. Close, USN
USS *Gurke* (DD-783)   Comdr. Frederick M. Radel, USN
USS *Henderson* (DD-785)   Comdr. William S. Stewart, USN[13]

Shortly before noon, *Mansfield* (the lead destroyer) reported what appeared to be a string of mines. The order was given to open fire and both the cruisers and destroyers began shooting at the enemy mines, with *Gurke* scoring the first hit. Fortunately, owing to the decision to proceed up the channel at low tide, the mines, being exposed, were discovered, and most were destroyed by gunfire. *Henderson*, the last ship in the column, was temporarily detached to destroy as many mines as possible before the rising flood tide again hid them from view.[14]

The other destroyers continued on toward Wolmi-do, moving through the harbor to their assigned positions, some less than half a mile from the fortified island. The ships anchored at short stay, with their bows swung southward into the flooding current. Shortly before 1300 the five destroyers commenced fire on the island's batteries and on the Inchon waterfront. After several minutes of undisturbed shore bombardment, the enemy batteries opened up.[15]

Their fire was concentrated on *Swenson*, *Collett*, and *Gurke*, as they were nearest the island and, over a period of twenty minutes, hits were made on all three. *Collett* took nine 75-millimeter hits, one of which disabled her computer and forced her to fire in local control. Three hits were made on *Gurke*, and fragments from a near miss on the *Swenson* killed an officer aboard. The engagement continued for nearly an hour until, at 1347, the destroyers weighed anchor and proceeded down channel. Casualties suffered were one killed and five wounded.[16]

The reconnaissance in force had been intended to draw the fire of North Korean batteries, in order that their location would be revealed for neutralization by destroyer or cruiser gunfire, or by air strikes. The bombardment was highly successful, and the press, in acknowledgement of the danger the ships had faced, dubbed the units of the Destroyer Element, 'Sitting Ducks.' Upon retirement from the harbor, some of the enemy guns that had not been silenced targeted the destroyers as they exited through a narrow channel. This was when an officer was killed, and another wounded, aboard *Swenson*.[17]

Photo 6-2

Wolmi-do (Island) under bombardment, on 13 September 1950, two days before the landings at Inchon. Sowolmi-do, connected to Wolmi-do by a causeway, is at the right, with Inchon beyond. A 40mm gun mount aboard the destroyer USS *Lyman K. Swenson* is in the foreground. She had earlier destroyed a floating mine with 40mm gunfire. National Archives photograph #80-G-420044

As soon as the destroyers were clear, the cruisers opened fire from the lower harbor against the Wolmi batteries. Shooting lasted until 1640, with one intermission for an air strike against the enemy, and then the task group retired seaward. All six ships of the Destroyer Element were subsequently awarded the Navy Unit Commendation.[18]

The following day, *Collett* was detached because of her damage, and that morning the remaining four destroyers of the initial bombardment group, rejoined by *Henderson* and joined and supported by the cruisers, once again stood (proceeded) back up the channel. Little resistance to their bombardment was offered this time. In mid-afternoon, with Wolmi-do ready for the Marines, the ships retired and then rendezvoused with other units of the Attack Force preparatory to the next day's assault landings.[19]

SCREENING AND PROTECTIVE GROUP (TG 90.7)

While the cruisers and destroyers of the Gunfire Support Group were knocking the 75mm shore batteries on Wolmi-do out of commission, the ships of a Screening and Protective Group were performing much less glamorous, but necessary, work.

Capt. John H. Unwin, DSC RN, commanding officer of the frigate HMS *Mounts Bay*, was responsible for forming an outer protective screen approximately 50 miles long, 40 miles south of the Inchon approaches. Its purpose was to prevent interference from enemy vessels, suicide craft, swimmers, floating mines, movement of the enemy between the islands and mainland, and to rescue any downed aircrews. All but two of the destroyers and frigates listed in the following table were under his command. The American destroyers *Rowan* and *Southerland* which were engaged in other duties were not part of the screen.[21]

Task Group 90.7 (Screening and Protective Group):
Capt. Richard T. Spofford, USN

Destroyers and Frigates	
USS *Rowan* (DD-782)	Comdr. Alan R. Josephson, USN
USS *Southerland* (DDR-743)	Comdr. Homer E. Conrad, USN
USS *Bayonne* (PF-21)	Lt. Comdr. Harry A. Clark, USN
USS *Newport* (PF-27)	Lt. Comdr. Percy A. Lilly Jr., USN
USS *Evansville* (PF-70)	Lt. Comdr. Eliot V. Converse Jr., USN
HMS *Mounts Bay* (F627)	Capt. John Henry Unwin, DSC RN
HMS *Whitesand Bay* (F633)	Lt. Comdr. J. V. Brothers, RN
HMNZS *Tutira* (F517)	Lt. Comdr. Peter J. H. Hoare, RN
HMNZS *Pukaki* (F424)	Lt. Comdr. Laurance E. Herrick, DSC RN
FS La *Grandiere* (F731)	Comdr. Urbain E. Cabanie, FN
Minesweepers	
USS *Pledge* (AM-277)	Lt. Richard Young, USN
USS *Partridge* (AMS-31)	Lt. (jg) Robert C. Fuller Jr., USN
USS *Mockingbird* (AMS-27)	Lt. (jg) Stanley P. Gary, USN
USS *Kite* (AMS-22)	Lt. (jg) Nicholas Grkovic, USN
USS *Osprey* (AMS-28)	Lt. (jg) Philip Levin, USN
USS *Redhead* (AMS-34)	Lt. (jg) T. R. Howard, USN
USS *Chatterer* (AMS-40)	Lt. (jg) James P. McMahon, USN[20]

A description of the activities of the minesweepers (a component of Task Group 90.7) and, like the destroyers and frigates, under the command of Capt. Richard T. Spofford, follows that of Task Force 91.

Admiral Andrewes commanded Task Force 91 (the Blockade and Covering Force), which consisted of the light carrier HMS *Triumph*, cruiser HMS *Ceylon*, eight destroyers and 15 ROK Navy vessels. This force was responsible for conducting special reconnaissance and covering missions prior to D-Day; providing cover for the vessels of the attacking force en route to Inchon; maintaining a naval blockade of the Korean west coast, south of 39°35' North; and performing such interdiction missions as might be assigned.[22]

Task Force 91 (Blockade and Covering Force):
Rear Adm. William G. Andrewes, RN

HMS *Triumph* (R16)	Capt. Arthur David Torlesse, DSO RN
HMS *Ceylon* (C30)	Capt. Cromwell F. J. L. Davies, DSC RN
HMS *Cockade* (D34)	Lt. Comdr. Herbert Jack Lee, DSC RN
HMS *Charity* (D29)	Lt. Comdr. Peter R. G. Worth, DSC RN
HMCS *Cayuga* (DDE218)	Capt. Jeffery V. Brock, DSC RCN
HMCS *Sioux* (DDE225)	Comdr. Paul Dalrymple Taylor, RCN
HMCS *Athabaskan* (DDE219)	Comdr. Robert Phillip Welland, DSC RCN
HMAS *Bataan* (D191)	Comdr. William B. M. Marks, RAN
HMAS *Warramunga* (D123)	Comdr. Otto H. Becher, DSC RAN
HNMS *Evertsen* (D802)	Lt. Comdr. D. J. Van Doorninck, RNN

ROK Naval Forces

Paik Doo San (PC-701)	Comdr. Chai Yong Nam, ROKN
Kum Kang San (PC-702)	Comdr. Lee Hi Jong, ROKN
Sam Kak San (PC-703)	Comdr. Lee Hung So, ROKN
Chiri San (PC-704)	Lt. Comdr. Hyun Sibak, ROKN
Tongyeong (JMS-302)	
Daegu (JMS-303)	
Danyang (JMS-306)	
Dancheon (JMS-307)	
Kang Jim (YMS-501)	
Kyong Chu (YMS-502)	
Kwang Chu (YMS-503)	
Ganggyeong (YMS-510)	
Guwolsan (YMS-512)	
Ko Yung (YMS-515)	
Yong Kung (YMS-518)[23]	

To simplify the conduct of his blockade and escort assignments, Andrewes split his forces into a Northern Group (TG 91.1) and a Southern Group (TG 91.2). To the Canadian ships fell the somewhat mundane duties of Southern Group. Capt. Jeffery V. Brock, DSC RCN, its commander, was responsible for:

- Providing escort for the logistic support group supplying the attacking force
- Enforcing a blockade of the coast between 35°45' and 36°45' North
- Maintaining a hunter-killer group to deal with enemy submarines in the unlikely event that they made their appearance in the area[24]

In execution of these duties, Brock had available the three Canadian destroyers and a few light ROK naval vessels. These ships carried out their duties with their expected efficiency and dispatch, but encountered no opposition from the enemy to the passage of the logistic support ships, nor found any hostile submarines to pursue.²⁵

During the period 12-21 September, five destroyers—HMAS *Warramunga* and HM ships *Charity*, *Cockade*, and *Concord*—were assigned duties as screening ships for the light carrier HMS *Triumph*. HMS *Concord* (D63) had just arrived in theater earlier that month. The senior officer of this group was Comdr. Otto Humphrey Becher, DSC RAN, the commanding officer of *Warramunga*.²⁶

Photo 6-3

Capt. Otto H. Becher, RAN, on the bridge of *Warramunga*, during patrols off the Korean coast, circa February 1951.
Australian War Memorial photograph DUKJ3812

MINE SQUADRON THREE FINDS LITTLE WORK

During pre-invasion bombardment of artillery sites, the Gunfire Support Group apparently found and destroyed all, or most, of the enemy-laid mines in the Inchon Channel. When the units of Mine Squadron 3 began sweeping the inner anchorages of Inchon Harbor at 0600 on D-Day (15 September), they found no mines and withdrew that afternoon.²⁷

Comprising the squadron were the USS *Pledge* (AM-277), *Partridge* (AMS-31), *Mockingbird* (AMS-27), *Kite* (AMS-22), *Osprey* (AMS-28), *Redhead* (AMS-34), and *Chatterer* (AMS-40). The flagship *Pledge* was a 184-foot, *Admirable*-class steel-hulled minesweeper; the remainder smaller wooden-hulled vessels, former YMSs reclassified as auxiliary minesweepers.[28]

Photo 6-4

From left to right, four auxiliary minesweepers of Mine Division 31—USS *Merganser* (AMS-26); USS *Osprey* (AMS-28); USS *Chatterer* (AMS-40) and USS *Mockingbird* (AMS-27)—nested together at Yokosuka, Japan, 30 November 1950. National Archives photograph #80-G-424597

1ST AND 5TH MARINE DIVISIONS LAND AT INCHON

The invasion of Inchon began at 1730 the evening of 15 September. The Marine Corps, for the first time in its history, had to scale seawalls before beginning an assault into the heart of a large city. These walls added challenges to the prospect of heavy opposition from an enemy hidden in warehouses, buildings, and other cover.[29]

The main landings, late in the day, followed the capture that morning of Wolmi-do, by the 3rd Battalion, 5th Marines. Both the island assault and that at Inchon were preceded by shore bombardment and strikes by carrier aircraft loaded with combinations of bombs, rockets, and napalm.[30]

Within the city, fighting continued through the hours of darkness, until midnight when the landing force reached its initial objectives. At this time, the 5th Marines controlled the hills above Red Beach, and thus the source of logistic build-up and supply. They had also advanced southward as far as the tidal basin, while the 1st Marines had reached the high ground north of Blue Beach, commanding the main road to Seoul. The cost of D-Day was 174 U.S. casualties, including 20 killed in action, 1 missing, and 1 who later died of his wounds.[31]

Photo 6-5

1st Lt. Baldomero Lopez, USMC, leads the 3rd Platoon, Company A, 1st Battalion, 5th Marines over the seawall on the northern side of Red Beach, 15 September 1950. Wooden scaling ladders are in use. Lopez was killed in action within a few minutes, while assaulting a North Korean bunker.
Naval History and Heritage Command photograph #NH 96876

That night, both the 1st and 5th Marines received orders directing them to attack after dawn, and move forward out of Inchon as quickly as possible. The heart of the seaport was left to be cleared by the 1st Korean Marine Corps (KMC) Regiment of 2,786 men strong.[32]

LEST WE FORGET

Photo 6-6

A chaplain reads the Last Rites service as Lt. (jg) David H. Swenson Jr., USN, is buried at sea from the cruiser USS *Toledo* (CA-133), off Inchon, Korea. The destroyer USS *Lyman K. Swenson* (DD-729) is in the background, with her crew at quarters on deck. Naval History and Heritage Command photograph #NH 96980

Lt. (jg) David H. Swenson Jr., USN, the officer killed aboard the destroyer USS *Lyman K. Swenson* (DD-729), was awarded the Silver Star Medal (posthumously) for conspicuous gallantry and intrepidity in action against the enemy. His citation reads in part:

> Lieutenant, Junior Grade, Swenson, while assigned the duties of Gunnery Liaison Officer, was charged with ascertaining from available information on hand the correct and specific targets on which his ship should fire during its bombardment of Inchon, Korea, on 13 September 1950, and to keep the ship's gunner control officer informed accordingly. While carrying out his assigned duties of observing the effect of his ship's gunfire on the enemy shore batteries and applying that information to the ship's gunfire charts in order that the guns could receive accurate revised target data, Lieutenant, Junior Grade, Swenson was struck by enemy counter-battery fire and instantly killed. By his courageous action and devotion to duty in refusing to leave his unprotected post in the face of heavy enemy fire, Lieutenant, Junior Grade, Swenson was responsible for obtaining and furnishing such valuable information to the gunnery control officer that he definitely assisted his ship to escape damage from enemy gunfire, thereby providing a material contribution to the war effort of the Korean Campaign.[33]

7

Autumn 1950, Wonsan and Chinnampo

Photo 7-1

Post-war view of Kunsan looking seaward, 1954. The many tents, houses, and maintenance buildings spread over the area are part of a joint USAF/RAAF base. Australian War Memorial photograph P02948.075

On 25 September 1950, following the Inchon landings, the Blockade and Covering Force (Task Force 91) was dissolved, and Admiral Andrewes' naval force resumed its Task Group 95.1 role, blockading duties along the west coast of Korea. From early July 1950 through the end of the war in July 1953, responsibility for the west coast naval blockade mission was assigned to the Royal Navy and her attached British Commonwealth and other Allied naval forces, including those of Australia, Canada, Colombia, France, the Netherlands, and New Zealand.[1]

The American light carrier USS *Bataan*, and escort carriers USS *Badoeng Strait*, USS *Barioko*, USS *Rendova*, and USS *Sicily*, while operating in the Yellow Sea and attached to the British command, alternated duty with the British and Australian carriers—HMS *Triumph*, HMS *Theseus*, HMS *Ocean*, HMS *Glory*, and HMAS *Sydney*. Although not under direct British command, several Republic of Korea naval vessels also supported the western naval forces' operations.[2]

Experience had already shown that, owing to the geographical characteristics of the west coast—high rise and fall of tide, very shallow water near the coastline, and a large number of islands—close inshore blockade activities by large naval vessels were almost impossible. These challenges posed by restrictive coastal waters were intensified by enemy usage of sea mines, from early September 1950 onward.[3]

On 7 September, the cruiser HMS *Jamaica* reported a number of mines near the sea approaches to Chinnampo, and on the next day, a second cruiser, HMS *Ceylon* also reported similar types of mines off Fankochi Point, northwest of Haeju. These mines were of a type that was easily laid by North Korean shallow-draft sampans safely out of the reach of UN ships. The threat of mines, then revealed, restricted the movements of ships of the Blockade Force, shrinking their patrol areas.[4]

MINE CLEARANCE AT KUSAN ON WEST COAST

Photo 7-2

Royal Canadian Navy destroyer HMCS *Athabaskan* during the Korea War. Australian War Memorial photograph P05890.060

As would soon become apparent, other areas were also mined, including Kusan, located 120 miles southwest of Seoul on the south bank of the Geum River just upstream from where it flowed into the Yellow Sea. On 25 September 1950, the Royal Canadian destroyer HMCS *Athabaskan*, and the ROKN vessels submarine chaser *Chiri San* (PC-704) and minesweeper *Danyang* (JMS-306) reconnoitered two small islands at the entrance to Kunsan harbor—Piun-to and Youjiku-to—which were believed to be held only by weak Communist police forces.[5]

In the first action to investigate possible enemy occupation of the islands, *Athabaskan* sent in her two motor cutters that morning with a party of 30 specially trained volunteers (led by Lt. Comdr. Thomas S. R. Peacock, RCN) to Piun-to. There was no opposition to the landing, and finding no signs of military presence, the party returned aboard ship.[6]

That afternoon, as boats from *Chiri San* and *Danyang* approached Youjiku-to under covering fire from *Athabaskan* and their own ships, they were met by machine-gun fire. During the ensuing exchange of gunfire, the enemy withdrew inland. Determining that the island was occupied by Communist troops, and not being prepared to take and hold the island against strong opposition, the South Koreans withdrew.[7]

While the ROK ships were supporting their landing parties, the *Chiri San* sighted a mine off the island. At low tide the following morning, *Athabaskan* sent in her motor cutters and a 14-foot dinghy, to investigate the sighting. An exposed moored mine was found at once, and marked with a dan buoy for later examination, while a second was sunk with rifle fire. Two others were sighted and their locations plotted on a chart before the rising tide hid them from view.[8]

The following day, a plan was devised to deal with the mines. It had proved difficult to sink the sea-emplaced ordnance with rifle fire, and it would be unwise for *Athabaskan* to venture close enough inshore to use her 40mm guns to destroy those remaining. The solution conceived was unorthodox; one that would require iron nerves, steady hands and expert boat handling. On 27 September, low tide was at 0900 and, under these conditions, one or more motor cutters and a dinghy set to work. A cutter went in towing the dinghy, and as mines were sighted, the dinghy, carrying Commissioned Gunner David William Hurl, RCN (the *Athabaskan*'s demolition expert) and others in his party, was released from the cutter and rowed to the spot.[9]

In turn, carefully avoiding a collision with the protruding horns of each contact mine (which, if detonated, would obliterate the boat and its occupants), the oarsman would bring the dinghy close alongside each respective mine and hold it there while Hurl and his assistants fastened time-fused demolition charges to the mine's mooring rings. After the fuses were lit, the dinghy would rapidly row clear, and remain at a safe distance until the charges exploded. Then it was back on task, working as quickly as prudent to render safe some more, before the tide came in hiding the locations of these menacing devices.[10]

Photo 7-3

Demolition Party from HMCS *Athabaskan* attaches a charge to a mine. Standing up is AB (Able Seaman) David Kidd, with oars is AB Ron Souliere, at right in boat is Petty Officer Tom Shields. Leaning over mine is Commissioned Gunner David Hurl, while AB Edward Dalton holds the deadly enemy ordnance steady.
Lt. Piosz/Canada Dept. of National Defence Library and Archives Canada PA-176294

In some cases, the charges merely blew holes in a casing, causing the mine to sink, while in others, there might be a tremendous explosion if the explosive material within the mine was set off. Working in this fashion, Hurl and his party were able to neutralize four mines before the tide forced a suspension of operations.[11]

Hurl was awarded a Mention in Despatches, and Petty Officer Second Class Thomas Shields, RCN, the British Empire Medal, for their personal courage. Shields' medal citation reads:

> Petty Officer Thomas Shields was one of a party which succeeded in destroying five enemy laid moored mines off Kunsan West Korea on the 26th and 27th September 1950. Petty Officer Shields prepared the detonation charge necessary to effect the mines destruction, and then assisted in securing the charge to the mines. This work was carried out under difficult conditions with a fast tide running, making the management of the mines and their small boat uncertain during the critical operation of attaching the charge. In this operation, Petty Officer Shields displayed high personal courage and professional skills.[12]

Mention in Despatches

British Empire Medal

A member of the Armed Forces Mentioned in Despatches (MiD) is one whose name appears in an official report written by a superior officer and sent to higher command, in which his or her gallant or meritorious action in the face of the enemy is described. Service members of the British Empire or the Commonwealth mentioned in despatches are not awarded a medal for their actions, but wear an oak leaf device on the appropriate campaign medal. The British Empire medal is awarded for meritorious civil or military service worthy of recognition by the Crown.

The Australian destroyer HMAS *Bataan* joined *Athabaskan* on the inshore patrol at noon that day, 27 September. After receiving a report that the Communists, alarmed by the ROK landing two days earlier, had reinforced Youjiku-to with several hundred fresh troops, the destroyers joined forces to shell the island. With *Athabaskan*'s ordnance officer, Lt. Clarence A. Sturgeon, spotting from a boat close inshore and directing the fire of the destroyers by voice radio, the bombardment was deadly accurate, and systematically destroyed the targets taken under fire. HMAS *Bataan* remained in the Kunsan area that night, while HMCS *Athabaskan* moved north to the vicinity of Ammin to monitor enemy coastal traffic.[12]

As was the usual expectation at that time, there was nothing moving, so in the morning (28 September), *Athabaskan* returned to Youjiku for more mine demolition work. *Bataan*'s boats also joined in the work, destroying another four mines and finding a new minefield. The two ships joined forces in the afternoon for a bombardment of the Beijaa Bay area that HMCS *Cayuga* and *Athabaskan* had shelled early on the 22nd. After the bombardment, *Athabaskan* sailed north again to her night patrol area. Another uneventful night passed searching for enemy junks off Ammin-to, and *Athabaskan* departed on the morning of 29

September for Inchon. There, she turned over her duties to HMAS *Warramunga*, who relieved her on the Kunsan patrol.[13]

After taking on fuel, *Athabaskan* embarked the commander of the blockade forces on the west coast, Admiral Andrewes, for passage to Japan. Andrewes was not pleased with the then current situation, as expressed by a later comment, "I was disturbed lest the British and Commonwealth forces should have been given a purely holding role on the west coast." He believed the importance of the west coast blockade had significantly lessened since the Inchon landing, and that it did not make sense to subject the blockade to the potential risk of mines. During the first year of the war, the west coast was regarded as a relatively less important area than the east.[14]

The western side of North Korea was mainly an agricultural area which had little industry, and there were no railways or main supply routes within naval gunfire range.[15]

Conditions were markedly different on the east coast. Owing to its steep coastline, few islands offshore, and the fact that North Korea's main road and rail lines ran along the coast, naval bombardment was very advantageous. Results achieved, however, were usually short-term. Due to the enemy's unlimited manpower to keep communication lines operating, and aggressive use of mines, there was an ongoing need for naval blockade, bombardment, and minesweeping.[16]

MINESWEEPING AT KUNSAN

Photo 7-4

At left: sailors handling a float off the side of a minesweeper, probably the USS *Mockingbird* (AMS-27). The float is used to keep the sweep gear, once streamed astern, from striking the sea floor. At right: a sailor attaches a mechanical cutter to a sweep wire, which a float will keep from "bottoming out." The moor of a mine that encounters a wire will ride along it, and be severed by one of the cutters affixed to it.
National Archives photographs #80-G-425998 and #80-G-426011

On 2 October, the destroyers HMAS *Warramunga* and HMAS *Bataan* provided close support for a handful of U.S. Navy minesweepers, sent to clear the approaches to Kunsan. (*Bataan* left later that day for Sasebo.) Over the course of two days, 2-3 October, the minesweeper *Pledge* and four auxiliary minesweepers—*Mockingbird*, *Chatterer*, *Kite*, and *Redhead*—swept a 1,500-yard-wide, 22-mile-long channel into the port. In the process, they destroyed seven mines.[17]

UN GROUND FORCES ADVANCE NORTHWARD

Following the landings at Inchon then behind enemy lines, UN forces made rapid progress northward while pursuing fleeing enemy troops. The U.S. 1st Marines recaptured Seoul on 24 September and, only a little more than a week later, General MacArthur issued a surrender ultimatum to North Korean forces, on 1 October 1950. South Korean military forces crossed the 38th parallel that same day. On 7 October, the UN General Assembly passed a resolution which, although clearly not calling for the conquest and occupation of North Korea, gave implicit assent for combat operations by UN forces north of the 38th parallel.[18]

By 13 October, linked to the advance of the UN troops toward Haeju (69 miles south of Pyongyang the North Korean capital city), it was clear that the majority of the Communists had evacuated from the Haeju-Ongjin area. On the 19th, South Korean and US troops captured Pyongyang but, on that same day, Chinese troops began crossing the Yalu River into North Korea, largely undetected by the UN Command.[19]

BLOCKADE FORCES SHIFTED TO EAST COAST

With the rapid progress of UN forces northward, maintaining the west coast blockade patrol became of lesser importance to Allied navies. As a consequence, Admiral Andrewes detailed the cruiser HMS *Ceylon* and three destroyers—HMS *Cockade*, HMCS *Athabaskan*, and HMAS *Warramunga*—to work with the Gunfire Support Group on the east coast. Subsequently, five frigates (the British HMS *Mounts Bay*, New Zealander HMNZS *Pukaki* and *Tutira*, and French ship *La Grandiere*) were directed to join the resources of the Minesweeping and Protection Group under Capt. Richard T. Spofford, USN. As a result, only nine ships remained on the west coast for blockade duty, including the British light carrier HMS *Theseus*, which had recently relieved HMS *Triumph*.[20]

By late October, with news that lead elements of the ROK 1st Division had occupied the border town of Ch'osan on the Yalu River, the UN Command deemed it time to reduced naval forces. As a consequence, the aircraft carriers USS *Valley Forge* and USS *Boxer* departed their operating areas and, on 15 November, Admiral Joy authorized Admiral Andrewes to reduce the British Commonwealth naval forces. Departing ships were ordered to Hong Kong and Japan, to allow for emergency contingencies. They were to be kept at short notice if a quick recall was necessitated.[21]

Map 7-1

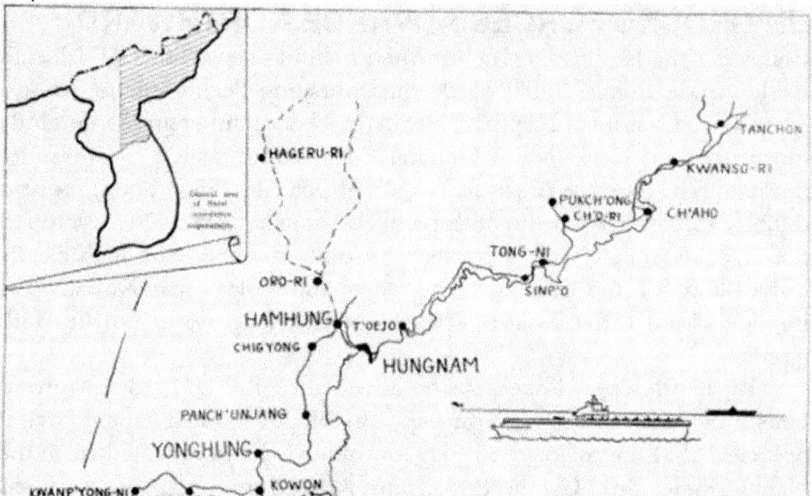

Area of naval air and shore bombardment interdiction on Korea's northeast coast
Navy Interdiction Korea Vol. II pamphlet, Rear Adm. Combs

RETIREMENT OF SHIPS FROM THEATER

This reduction of west coast blockade ships began with the destroyer HMS *Constance*, which left Sasebo for Hong Kong, on 20 November, followed by other Commonwealth ships under Andrewes' command. The admiral sailed for Hong Kong, on 25 November, in the cruiser HMS *Kenya*, after relinquishing command of Task Group 95.1 to Capt. Cromwell F. J. Lloyd-Davies, commanding officer of the light cruiser HMS *Ceylon*.[22]

Under the command of Lloyd-Davies, TG 95.1 was comprised of seven destroyers and four frigates. These ships, with the then optimistic expectation that the war would be over by Christmas, continued routine blockade patrols on the west coast.

Task Group 95.1: Capt. Cromwell F. J. Lloyd-Davies, RN

Destroyers	Destroyers	Frigates
HMAS *Bataan*	HMCS *Sioux*	HMS *Cardigan Bay*
HMAS *Warramunga*	HMS *Cossack*	HMS *Morecambe Bay*
HMCS *Athabaskan*	HNMS *Evertsen*	HMNZS *Rotoiti*
HMCS *Cayuga*		HMNZS *Tutira*[23]

LANDINGS AT WONSAN

Intervening events in autumn 1950, before the reduction in size of the west coast Blockading Force, included the landing of Tenth Corps, commanded by Maj. Gen. Edward Almond, on Korea's northeast coast at Wonsan, and mine clearance operations at Chinnampo (the port of Pyongyang) on the west coast.

Believing that his ultimate military objective was the destruction of the North Korean military forces, MacArthur planned that Lt. Gen. Walton Walker's Eighth Army would attack across the 38th parallel, concentrating its main effort along the Kaesong-Sariwon-Pyongyang axis while Almond's USMC Tenth Corps were landing at Wonsan. Almond would then move northward between the Sea of Japan and the Taebek Mountain Range, turning westward through passes in the mountains to link up with Walker' Army and thereby trapping the remnants of the North Koreans. The plan envisioned that these two commands, after uniting, would advance north to the Chongju-Kunuri-Wongwon-Hamhung-Hungnam line, which stretched a mere fifty to one hundred miles south of the Yalu River. The river marked the border between Korea and Red China.[24]

MacArthur's decisive plan made two important assumptions: first, that the bulk of the North Korean Army had already been destroyed, and second, that neither the USSR nor Red China would enter the conflict. The decision to land troops at Wonsan resulted in the loss of

Allied minesweepers to Soviet-supplied mines during clearance operations to open the port, and delayed planned operations ashore, and accordingly received much criticism.[25]

Photo 7-5

Amphibious shipping anchored in Wonsan's outer harbor during the landing of the 1st Marine Division, 26 October 1950. During mine clearance operations to open the harbor prior to troop landings, four minesweepers—USS *Pirate* (AM-275), USS *Pledge* (AM-77), ROK *Gongju* (YMS-516), and Japanese minesweeper *No. 14*—were lost, with associated large numbers of personnel casualties.
National Archives photograph #80-G-422091

MacArthur later provided an explanation of the factors necessitating a seaborne landing at Wonsan:

> The Eighth Army's lines of supply were already taxed to their maximum capacity to sustain day-to-day minimum requirements of its troops in the line. Furthermore, the dispatch of Tenth Corps by sea was intended as a flanking movement against enemy remnants still trying to escape from the south to the north, and as an envelopment to bring pressure upon Pyongyang should the attack upon that enemy capital result in a long drawn out siege.[26]

Opening the port of Wonsan, and maintaining it open, required considerable effort by minesweepers and surface combatant ships over the course of the war. North Korea, desperate to deny the Allies use of the port, kept mining its waters, and taking under fire with artillery, minesweepers plying their trade. The openings in caves in which shore guns were hidden, were disguised by foliage part of the year, and white cloth in winter months to match the area sheathed in snow. Accordingly, destroyers tasked with providing protection for the small, wooden-hulled minesweepers working near shore, had to react quickly to muzzle flashes, maintaining "top form" to destroy with counterfire, the enemy mobile guns which were easily rolled out, and rolled back into hillside caverns.

Photo 7-6

Drawing by Hugh Cabot of the destroyer USS *Gregory* (DD-802), 1952, patrolling inside Wonsan Harbor. Carefully concealed shore batteries and bunkered artillery required expert observation from destroyer gunner's mates, and involved a high degree of risk in detecting gun positions and eliminating them.
Naval History and Heritage Command photograph #88-187-W

MINESWEEPERS EARN HIGHEST US MILITARY UNIT AWARD FOR THEIR HEROISM AT WONSAN, KOREA

The ten U.S. Navy minesweepers identified in the table were awarded the Presidential Unit Citation (PUC) for heroism of their crews at Wonsan. The PUC is awarded to military units of the United States, and those of Allied countries, for extraordinary heroism in action against an armed enemy. The collective degree of valor against an armed enemy

72   Chapter 7

by a recipient unit is the same as that required for award of the Navy Cross to an individual.

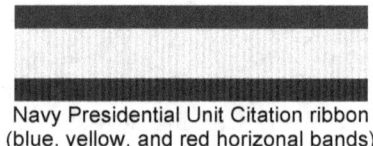

Navy Presidential Unit Citation ribbon
(blue, yellow, and red horizonal bands)

Incredible (AM-249)	10-24 Oct 50	*Merganser* (AMS-26)	11-24 Oct 50
Pirate (AM-275)	11-12 Oct 50	*Mockingbird* (AMS-27)	10-24 Oct 50
Pledge (AM-277)	10-12 Oct 50	*Osprey* (AMS-28)	10-24 Oct 50
Chatterer (AMS-40)	10-24 Oct 50	*Partridge* (AMS-31)	10-24 Oct 50
Kite (AMS-22)	10-24 Oct 50	*Redhead* (AMS-34)	11-24 Oct 50

OPENING OF CHINNAMPO ("NAMPO")

Photo 7-7

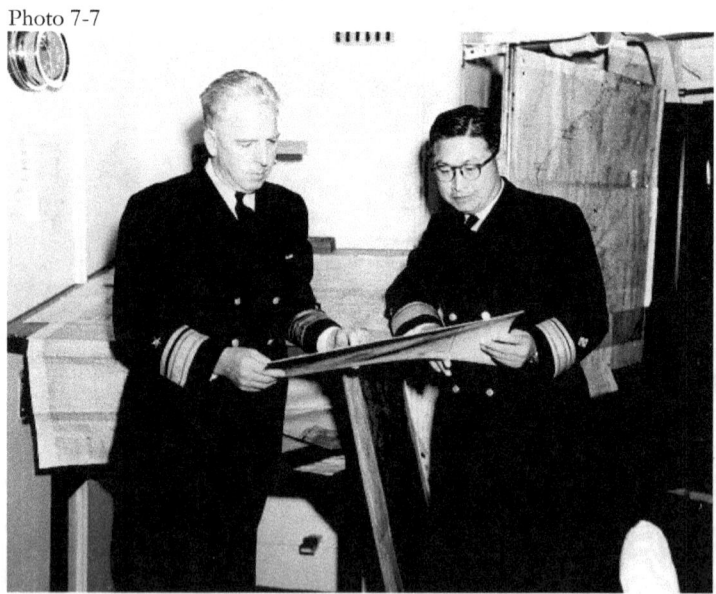

Rear Adm. Allan E. Smith, USN (left), commander, Task Force 95, discusses Korean War operations with Rear Adm. Won Il Sohn, Republic of Korea chief of Naval Operations, on board the destroyer tender USS *Dixie* (AD-14), 6 December 1950. National Archives photograph #80-G-423956

Even before Wonsan was opened on the east coast, on 25 October 1950, Admiral Joy had ordered the establishment of a mine clearance force for Chinnampo on the west coast. Following movement by United States and ROK forces northward along the Korean west coast, and capture of Pyongyang, on 19 October, the supplies flowing in to Inchon for this front were too far from the combat zone. The increasing distance from the front and the relatively small capacity of the Inchon port facilities made it imperative that Chinnampo be made available to support the U.S. Eighth Army.[27]

Ordered to sweep into Chinnampo by commander, Far East Naval Force, three days before a channel was finally cleared through to the beaches of Wonsan, Rear Adm. Allan E. Smith (commander, TF 95) had neither the people nor the ships to accomplish the task. All the minesweepers in Korea were already committed to mine clearance at Wonsan and Iwon. (The U.S. Seventh Army division landed at Iwon, located north-northeast of Wonsan, on 29 October, after the destroyer minesweepers USS *Doyle* (DMS-34) and *Endicott* (DMS-35) found no traces of mines in the harbor.)[28]

Undeterred, Smith dispatched two mine warfare experts to plan and execute the operations. He sent Comdr. Donald N. Clay (assigned to the commander, Pacific Fleet staff) to Chinnampo in search of mine intelligence, and Comdr. Stephen Morris Archer (Mine Forces, Atlantic Fleet) to Sasebo to commandeer any ships he could find there.[29]

Archer, who would also be in charge of the sweeping operation, had extensive experience in mine warfare. He had been an observer aboard several Royal Navy minelayers during the winter, 1940-1941, and in 1941 he had assisted in the founding of the US Naval Mine Disposal Organization. From 1942 to 1943, Archer was in charge of naval mine and bomb disposal in the Office of the Chief of Naval Operations. Following these shore assignments, he commanded the destroyer USS *Power* (DD-839), and later Mine Division Four, in the Atlantic Fleet.[30]

Significantly, during his relatively short period in the Korean theater before his Chinnampo orders, Archer had earned a Silver Star, later awarded on 13 December 1950. His medal citation reads:

> The President of the United States of America takes pleasure in presenting the Silver Star to Commander Stephen Morris Archer, United States Navy, for conspicuous gallantry and intrepidity in action as Commander Underwater Reconnaissance Element in support of naval forces conducting operations in heavily mined waters during the period 10 to 22 October 1950. When the U.S.S. *Pledge* and U.S.S. *Pirate* were mined on 12 October, he conducted rescue operations for surviving personnel with disregard for his own safety in the face of enemy gunfire from shore batteries. The leadership, force, and judgment displayed by Commander Archer in directing visual and sonar searches for mines throughout this period and in supervising underwater demolition operations in the vicinity

of Koto and Rei-To Islands contributed directly to the successful clearance of mine channels and anchorage areas off Wonsan, Korea. His outstanding courage and steadfast devotion to duty were in keeping with the highest traditions of the United States Naval Service.[31]

AD HOC COLLECTION OF MINE CLEARANCE UNITS

Once assembled, Archer's mine countermeasures forces included: one destroyer; twenty-three minesweepers; a Japanese tank landing ship to serve as a helicopter platform; a dock landing ship carrying fourteen LCVP wooden-hulled landing craft ("Higgins boats") fitted with sweep gear for working in shallow waters; a high-speed transport with Underwater Demolition Team One embarked for mine disposal; and a salvage ship to provide assistance as required, should a minesweeper or other vessel be damaged.[32]

Fighter aircraft from the light carrier HMS *Theseus* provided protection overhead during minesweeping operations, and other type aircraft assisted in other ways. U.S. Navy PBM seaplanes and helicopters and Royal Navy Sunderland aircraft were used to spot mines, or to destroy them with their machine guns, and warn the sweepers.[33]

Task Element 95.69: Comdr. Stephen M. Archer, USN	
Type Ship or Unit	Name or Composition
destroyer	USS *Forrest Royal* (DD-872)
destroyer minesweepers	USS *Carmick* (DMS-33), USS *Thompson* (DMS-38)
auxiliary minesweepers	USS *Gull* (AMS-16), USS *Pelican* (AMS-32), USS *Swallow* (AMS-36)
ROK minesweepers	*Danyang* (JMS-306), *Kyong Chu* (YMS-502), *Kwang Chu* (YMS-503), *Kim Chon* (YMS-513)
Japanese minesweepers	*PS-56, PS-62, MS-03, MS-06, MS-08, MS-09, MS-10, MS-12, MS-13, MS-15, MS-21, MS-22, MS-23, MS-57*
Japanese tank landing ship	SCAJAP *LST-Q007* (for use as a helicopter platform)
dock landing ship	USS *Catamount* (LSD-17) with 14 minesweeping boats
high-speed transport	USS *Horace A. Bass* (APD-124) with UDT One
salvage ship	USS *Bolster* (ARS-38)
Supporting Aviation Assets	
RN light aircraft carrier	HMS *Theseus* providing fighter coverage
RAF seaplanes	Short Sunderland patrol bombers
USN seaplanes	PBM-5 Mariners aboard USS *Gardiners Bay* (AVP-39)
USN helicopter	pilot Lt. Robert D. Romer (perhaps from HU-1)[34]

The author is unsure of the exact type of helicopter Archer was able to obtain the use of, but it was likely a Sikorsky S-51 (HO3S-1) from U.S. Navy Helicopter Utility Squadron ONE (HU-1). Its aircraft had warned of mine lines ahead of the minesweepers at Wonsan. HU-1 would later receive the Presidential Unit Citation for rescuing downed airmen and sailors in the sea, directing naval gunfire against shore targets, and spotting sea mines in Korea.³⁵

Photo 7-8

Sikorsky HO3S-1 helicopter, of HU-1, 19 October 1950.
National Archives photograph 80-G-420949

Photo 7-9

A Royal Air Force Short Sunderland aircraft inside the seawall at Iwakuni, Japan, 1950.
Australian War Memorial photograph P00675.120

Photo 7-10

British light fleet aircraft carrier HMS *Theseus*, circa 1952.
Australian War Memorial photograph 302489

Mine clearing operations began, on 29 October, with *Thompson* and *Carmick* beginning to sweep in from the shallow Yellow Sea, 39 miles from the channel entrance. All ships were at work, by 2 November, with the helicopter spotting for mines ahead of the sweepers. *Catamount* and her boat sweepers arrived and went to work, on 5 November. Air search and patrol eased their toil considerably, reducing the danger that large quantities of mines posed to the minesweepers.[36]

Photo 7-11

A 36-foot, wooden-hulled LCVP landing craft ("Higgins boat") bucks in the well of the dock landing ship USS *Catamount* (LSD-17), during minesweeping off Chinnampo. The destroyer USS *Forrest Royal* (DD-872), flagship for the operation, is in the background.
National Archives photograph #80-G-422837

Martin PBM-5 Mariners based aboard the seaplane tender *Gardiners Bay* began searching Chinnampo waters, on 28 October, and were soon joined by RAF Sunderlands. In six weeks, the planes sighted 340 mines and disposed of many by machine gun fire, including six by PBM pilot Lt. Comdr. Randall T. Boyd Jr., in a mere half hour. To speed progress, Archer directed sweeping a clear path around the minefields whenever possible, leaving their mines to be dealt with later. As a result of these and other actions, no ships or personnel were lost in the entire operation—a remarkable achievement.[37]

Photo 7-12

USS *Gardiners Bay* (AVP-39) off Houghton, Washington, 18 February 1945. Fresh from her builder's yard, she is cloaked in camouflage paint. The seaplane tender would earn two battle stars in WWII, and another four in Korea.
National Archives photograph #19-N-85076

There were 212 mines found in the Chinnampo harbor itself. There would have been more, had aircraft from the *Theseus* not sunk what was believed at the time to be a mine carrying barge. Divers later located and searched the sunken barge, finding fifteen mines still aboard it. On 6 November, a Korean-manned tug with Comdr. Donald Clay on board, proceeded from Chinnampo to the Yellow Sea to prove that the channel was free of mines. He had assisted the mine clearance, considerably, by rounding up and "interrogating" a North Korean called "Shorty" who had helped to plant the mines, and knew where they all were planted. Next, the minesweeper *Kwang Chu* (YMS-503) entered the port from the sea, and finally, the arrival of the hospital ship USS *Repose* (AH-16), on the 20th, signaled Chinnampo's opening to larger ships.[38]

ANOTHER LAUREL FOR COMMANDER ARCHER

Comdr. Stephen Archer was awarded the Legion of Merit for his efforts in opening the Port of Chinnampo, under the very dangerous and difficult circumstances only alluded to in the preceding brief summary of the operation. Continuing his naval career after the war, he commanded the destroyer tender USS *Bryce Canyon* (AD36), and later Destroyer Flotilla Five. In addition to his Silver Star and Legion of Merit earned in Korea, Archer also had a Bronze Star from World War II, and the Order of the British Empire Medal.[39]

His Legion of Merit citation follows:

> The President of the United States of America takes pleasure in presenting the Legion of Merit with Combat "V" to Commander Stephen Morris Archer, United States Navy, for exceptionally meritorious conduct in the performance of outstanding services to the Government of the United States as Commander Task Element 95.69 during the sweeping of enemy laid mines in the approaches and harbor of Chinnampo, Korea, from 1 November to 22 November 1950. Discharging his many responsibilities with diligence and ability, Commander Archer organized and successfully directed the operations under the most difficult conditions due to the extremely bad weather, varied types of ships and limited personnel and equipment. He expeditiously swept the channel in the short time of ten days without the loss of a man or ship, thereby opening the port for supplies badly needed by our ground forces. By his judgment, outstanding professional skill and conscientious devotion to the fulfillment of an important task he contributed materially to the action against the enemy in Korea. Commander Archer's valiant and aggressive leadership was in keeping with the highest traditions of the United States Naval Service.

Photo 7-13

Capt. Stephen Archer, USN
Courtesy of NavSource

Silver Star

Legion of Merit

Bronze Star

Order of the British Empire

UNIT AWARDS FOR CHINNAMPO MINE CLEARANCE AND EARLIER UDT SPECIAL OPERATIONS

Navy Unit Commendation ribbon

USS *Horace A. Bass* (APD-124), Underwater Demolition Team One, and a group of Marine raiders also embarked aboard her, had earlier earned the same commendation for operations off the east coast of Korea. Led by Lt. Comdr. David Fife 'Kelly' Welch, USN, the Navy-Marine raiders destroyed three bridges, and reconnoitered possible landing beaches near Kunsan. Efforts to destroy railroad tunnels with explosives were unsuccessful—the use of piles of 60lb TNT satchels proved to be exercises in futility. Welch, who later retired in the rank of rear admiral, was awarded the Silver Star for these operations. His citation reads in part: "for conspicuous gallantry and intrepidity in action against the enemy while serving as Commanding Officer of Underwater Demolition Team ONE, a part of the raiding forces in a Special Operations Group of Amphibious Group ONE, Pacific Fleet which conducted a series of successful night demolition raids and beach reconnaissance missions in the coastal waters of enemy-held Korea during the period 12 through 25 August 1950."[40]

Navy Unit Commendation (Port of Chinnampo)

Ship or Unit	Dates
USS *Carmick* (DMS-33)	28 Oct-25 Nov 1950
USS *Thompson* (DMS-38)	28 Oct-25 Nov 1950
USS *Gull* (AMS-16)	2-23 Nov 1950
USS *Pelican* (AMS-32)	2-23 Nov 1950
USS *Swallow* (AMS-36)	2-23 Nov 1950
Minesweeping Boat Unit (TE 95.69)	29 Oct-29 Nov 1950

Navy Unit Commendation
(Special Operations Group of Amphibious Group One)

USS *Horace A. Bass* (APD-124)	12-25 Aug 1950
Reconnaissance Company	12-25 Aug 1950
Underwater Demolition Team One	12-25 Aug 1950

8

Evacuation of Chinnampo

The "Chinnampo affair" was without a doubt the most important mission performed by the Canadian Destroyer Division as a group during the entire Korean conflict. All the ships of Task Element 95.12 performed their duties ably, and most of them showed exceptional skill in accomplishing their difficult assignments. But it is to Cayuga, *as the senior ship, that most of the credit for the success of the operation must go. Captain Brock's handling of his task element was faultless and received the highest praise from his superiors.*

—Thor Thorgrimsson and E. C. Russell in *Canadian Naval Operations in Korean Waters 1950-1955*

Photo 8-1

Painting by Bo Hermanson of the destroyer HMCS *Cayuga* during the evacuation and subsequent shore bombardment of harbor facilities in the port of Chinnampo, Korea, on 5 December 1950, in order to deny these to the advancing North Korean Army. Her forward gun turrets are firing at the Chinnampo power generator station, in an effort to knock it out of action. The flames and smoke are shoreward of the ship.
Courtesy of Bo Hermanson

In late November 1950, it became apparent to the United Nations Command (UNC) that it had underestimated the strength of Chinese forces entering the war. Before conditions changed dramatically, U.S. Army 7th Infantry Division elements reached the Yalu River (near Hyesanjin in the northeast), on 21 November, the furthest Tenth Corps penetration of North Korea. Three days later, Eighth Army, located generally north of the Cheongcheon River in western North Korea, began a coordinated attack toward the Manchurian border.[1]

On the night of 26 November, the Chinese began attacks against Eighth Army forces that badly mauled ROK II Corps and the U.S. 2nd Infantry Division. The following day, the Chinese struck the 1st Marine Division west of the Jangjin (Chosin) Reservoir and the 31st Regimental Combat Team of the 7th Infantry Division east of the reservoir. On 28 November, Walker and Almond flew to Tokyo to meet with General MacArthur. Following this meeting, MacArthur sent a message to the Joint Chiefs of Staff, advising them that, with the intervention of the Chinese, the United States and the UNC faced "an entirely new war."[2]

On 30 November 1950, with the ground situation deteriorating, Rear Adm. James H. Doyle, USN, commander, Task Force 90, put all ships in port on two-hour notice and began to deploy them to Korea. Half of the task force, most ships of which were at Sasebo, were to make for the Korean west coast and come under Rear Adm. Lyman A. Thackrey, commander Western Redeployment Group (TG 90.1) and commander Amphibious Group Three. This force would be prepared to evacuate Eighth Army forces from Chinnampo and Inchon. The other half of Task Force 90 would proceed to the Wonsan–Hungnam area for the amphibious evacuation of Tenth Corps forces from Hungnam, Wonsan, and Seongjin in the northeast.[3]

EIGHTH ARMY REQUIRES IMMEDIATE ASSISTANCE

On 30 November and 1 December, the units of Transport Squadron One—attack transports *Bayfield*, *Bexar*, and *Okanogan*, and attack cargo ships *Algol* and *Montague*—sailed independently from Japanese ports. The squadron had been assigned to Task Group 90.1 and ordered to Inchon. On the afternoon of the 3rd, while heading northward into the Yellow Sea, Capt. Samuel G. Kelly (the squadron commander aboard *Bayfield*) intercepted a message from ComNavFE to CTG 90.1. It relayed an urgent Eighth Army request for the dispatch of Thackrey's ships to Chinnampo, and expressed Joy's doubts as to the possibility of loading and protecting so many large units there. Kelly continued to proceed northward, while awaiting amplifying instructions. At 2200, he decided to take positive action, by directing his dispersed units to join him off the Chinnampo swept channel in the morning.[4]

Photo 8-2

Attack transport USS *Bayfield* (APA-33), 6 February 1952, location unknown.
Naval History and Heritage Command photograph #NH 99243

A few hours later, at 0330 on 4 December, Kelly intercepted a message from Rear Adm. Allan E. Smith to Thackrey, reporting that the six west coast destroyers of TE 95.12 (the Surface Blockade and Patrol Element, then under Capt. Jeffrey V. Brock, RCN, in *Cayuga*) were available to protect the transports, and that the British cruiser *Ceylon* was starting from Sasebo for the west coast. Unknown to Kelly, additional assistance was also on the way. Admiral Andrewes, after a hasty return to Sasebo from Hong Kong, was preparing to sail with the light carrier HMS *Theseus* and four destroyers for the Yellow Sea.[5]

Photo 8-3

Destroyer escort USS *Foss* (DE-59), location and date unknown.
National Archives photograph #80-G-630973

Naval units then present at Chinnampo were the destroyer escort USS *Foss* (Lt. Comdr. Henry J. Ereckson), which was providing the city with electric power, and three motor launches located at a small Korean naval base. Off the mouth of the Taedong River, the Task Element 95.69 minesweepers were still at work. At 0236 on 4 December, Ereckson reported that the situation in Chinnampo was shaky; that he would keep the power on as long as possible; evacuate Eighth Army personnel; and at the last, if still senior officer, would form a convoy and get the shipping out. Japanese tank landing ships would take part in the evacuation. It is unclear, when these SCAJAP (Shipping Control Authority for the Japanese Merchant Marine) LSTs arrived at Chinnampo. The Korean base commander advised his superiors that Eighth Army had ordered him to redeploy at once and that, with 100 sailboats and 50,000 refugees trying to flee Chinnampo, he would try to send 30,000 out by sea and the remainder overland.[6]

ARRIVAL OF TRANSPORTS AT CHINNAMPO

Around noon on 4 December, following the arrival of the transports at Chinnampo (after a daytime, hazardous journey up the tortuous 84-mile channel, guided by minesweepers), Kelly assumed command of the evacuation. Ordering his ships to man their guns, lower their boats, and keep steam up to their engine throttles, he commenced loading. The only remaining personnel to be transported were the 1,700 soldiers of the 7th Medium Port and 501st Harbor Craft Platoon, which had its own shipping (various floating vessels, including tug boats, tankers, Q boats, and landing craft) and about 7,000 wounded ROK soldiers, government workers, military and civilian prisoners, and police.[7]

Finally, Kelly began to receive dispatches from his seniors: Brock's destroyers were heading his way; *Theseus* would have air cover there the next day; and he was in charge.[8]

"Force protection" would be most welcome. Quite apart from the hazards of navigation, Chinnampo was a poor place to be caught in, for the anchorage was within mortar range of the reverse slopes of the hills that fronted the harbor. Reports from Army personnel ashore indicated: an 80-mile gap in their lines to the north, the enemy was reported in Pyongyang and heading for Chinnampo, no combat forces were available, and the service troops manning the 'road blocks' were to be withdrawn at midnight. Captain Brock, in the midst of this situation, inquired as to the state of affairs and offered to bring his destroyers up the channel to Nampo in the dark, if needed. The offer was very welcome, and was immediately accepted.[9]

DESTROYERS RISK CHANNEL IN DARKNESS

Photo 8-4

Officers on watch on the open bridge of HMAS *Warramunga* during a patrol off the Korean coast. Lt. Rohan Edwin Lesh, RAN, is taking a bearing to plot the destroyer's position. Peering through binoculars is Sub Lt. Kenneth Malcolm Barnett, RAN. Australian War Memorial photograph DUKJ3820

Captain Brock in *Cayuga*, along with two other Canadian destroyers, had left Sasebo, on 20 November, to take over blockade duties in the coastal waters between Inchon and the mouth of the Yalu. With only one British cruiser in theater (the others being in Hong Kong), Brock was placed in command of the group, which other destroyers and ROK vessels would join, designated Task Element 95.12.[10]

At midnight, on 3 December, the element was maintaining its normal patrols, HMCS *Cayuga* and HMAS *Bataan* were operating near the mouth of the Yalu, and HMCS *Athabaskan* and HMAS *Warramunga* were patrolling to the south of them. HMCS *Sioux* was absent, escorting the British oiler RFA *Wave Laird* from Inchon to the element's replenishment area south of Cho-do. Three hours later, in early morning darkness on the 4th, Brock received orders from Andrewes:

> 5 APA transports have been diverted Chinnampo to withdraw some troops 8th Army.... Defence of these ships is now your primary aim. Blockade is secondary.... military situation serious. Be prepared to act in fire support of Eighth Army, entering Chinnampo River swept channel as necessary...[11]

At 1100, the arrival of the attack transports in the approaches to Chinnampo was reported and, shortly thereafter the destroyer USS *Forrest Royal* was assigned to reinforce TE 95.12. This addition was most welcome, as it enabled Brock to retain two destroyers on patrol while concentrating the remaining four in the replenishment area south of Cho-do (designated Area Shelter) to protect the transports.[12]

A message from Rear Admiral Smith (CTG 95), revealed that the situation had become serious and the transports would probably require actual gunfire support in Chinnampo Harbor itself, and not just covering support from destroyers positioned near Cho-do. There was a distinct possibility that the enemy might attack Chinnampo, and the six destroyers, including *Sioux* which had rejoined after completing her escort assignment, would then be required to enter the harbor to help in its defense. They would be joined later by *Ceylon*, expected to arrive, on the morning of 5 December, and take over command of the task element.[13]

In response to this situation, Brock ordered *Athabaskan*, *Sioux*, *Bataan*, and *Warramunga* to assemble near the entrance of the swept channel into Chinnampo, ready to proceed upriver in the morning. *Cayuga* was to remain at Area Shelter with *Forrest Royal* to await the arrival of *Ceylon*, the U.S. destroyer being retained to serve as the link with *Ceylon*, should *Cayuga* have to make the passage to Chinnampo before she arrived. With this plan in hand, but mindful that the military situation at Chinnampo was unknown, Brock sent a message to Kelly, requesting information and asking if the services of the destroyers would be required that night.[14]

Kelly's message confirmed that the presence of the destroyers was required and added, "The local situation may reach emergency basis Tuesday (5 December) forenoon." Responding to this information, Brock immediately ordered all six destroyers to prepare for a night passage to Chinnampo. He then recalled the ROKN vessels under his command—*Chiri San* (PC-704), *Daegu* (JMS-303), and *Toseong* (JMS-308)—from blockade patrol in the north. They were now to protect the seaward approaches to the Daido-ko from any attempted mining attacks during the absence of the destroyers.[15]

The minesweepers of Task Element 95.69 (minesweeping element), previously plying their trade off the Taedong River mouth, had been stationed at the critical turning points in the swept channel during transit of the attack transports, and their presence was to prove invaluable to the navigation of the destroyers as well.[16]

Cayuga left Area Shelter at 2230 and made for the swept channel, with *Athabaskan*, *Bataan*, *Forrest Royal*, and *Sioux* joining, in that order, to form a column astern of her. *Warramunga*, which had been anchored north of Cho-do near the entrance to the channel, weighed and set out ahead independently. The night was dark, the tide was almost at low ebb, and low water was the cause of the first casualty. *Warramunga* reported herself aground shortly before midnight. She worked herself free, returned to Area Shelter to examine her underwater hull and screws for damage, and took no further part in the channel transit.[17]

Sioux, the next casualty, went aground shortly after midnight, on a bank of sand and mud inside the marked channel. She was able to easily back clear but, while maneuvering, fouled her starboard screw. Unable, in darkness, to remove the mooring wire of an unlit buoy which had drifted into the channel, and was now entangled around her propeller, *Sioux* also returned to Area Shelter.[18]

Meanwhile, the four remaining destroyers continued to creep up the channel in darkness—a nerve-wracking journey for everyone, but particularly for commanding officers and navigators. In *Cayuga*, Lt. Andrew L. Collier, RCN, dashed back and forth between the radar screen and the chart table, rapidly and accurately plotting his "fixes," and relaying the ship's position and recommended rudder orders and courses to steer to Captain Brock, and information to the destroyers astern. Collier would later be awarded the Distinguished Service Cross for his actions, as described in an excerpt from his medal citation:

> The channel up the Daido-ko is narrow, tortuous and shallow plus had been heavily mined by the North Koreans. Lt. Collier was navigating officer on HMCS *Cayuga* on this dangerous passage at night. Collier made 132 fixes, most of them by radar, showing the position of the ship in relation to the channel marker buoys and nearby landmarks, and the accuracy of his navigation undoubtedly played a large part in ensuring the success of the entire operation to evacuate Chinnampo.[19]

The hazardous, lengthy channel-transit was finally over at 0330 on 5 December, when HMCS *Cayuga* anchored abreast of the main piers at Chinnampo; HMAS *Bataan* and USS *Forrest Royal* near the southern entrance to the harbor; and HMCS *Athabaskan*, upstream, north of the city. Captain Brock was annoyed to find, after the dangerous journey, "Chinnampo a blaze of lights and all peaceful and serene." However, the absence of North Korean troops was preferable to the alternative.[20]

At a conference that morning aboard USS *Bayfield*, Captain Kelly decided that he would devote his attention to the problems of loading and evacuation, and leave the arrangements for the defense of the harbor and of the evacuation forces in Captain Brock's hands. Upon the departure of Kelly with the transports, the command of all naval operations at Chinnampo would pass to Brock.[21]

Captains Kelly and Brock then met with Colonel Wilson (the U.S. Army commander at Chinnampo, whose men were engaged in loading ships) to discuss the final details of the evacuation. It was considered imperative that all the larger ships pass out the dangerous swept channel in daylight. Following them out before dark would be the attack transport USS *Bexar*, with the army demolition squads aboard her, and the destroyer escort USS *Foss* taking out the last of the naval personnel. Because of lack of time and a shortage of explosives, the army personnel would be unable to complete destruction of the huge stocks of military materiel left behind, and other infrastructure to deny its capture by enemy forces. Accordingly, the task was to be left to the destroyers.[22]

ACTIONS OF DESTROYERS

Small craft were scurrying back and forth between shore and ship, carrying wounded, prisoners of war, Korean civilian refugees and nonessential military personnel to the waiting transports and SCAJAP tank landing ships in the harbor. However, despite these hurried efforts, it became apparent by noon that the harbor could not be cleared of all shipping in time for the destroyers to carry out their bombardment and complete the navigation of the swept channel in daylight. In view of this, *Athabaskan* was sent five miles downriver, near the entrance to the swept channel, to establish a defended anchorage. At the conclusion of the bombardment, the other destroyers would retire to this anchorage, as an alternative to another hazardous night passage.[23]

When *Athabaskan* arrived at the site chosen for the anchorage, she quickly demolished three nearby "pill boxes" (a type of blockhouse, or concrete dug-in guard post) with her 4-inch guns. Her next action was to send off her motor cutters to search the innumerable junks and other civilian craft streaming downriver. A great many junks were boarded to ensure that the enemy was not endeavoring to mine the swept channel by this means, and thereby impede the evacuation. No sign of this activity was discovered.[24]

At 1400, the first two transports left the harbor, the forerunners of an exodus that continued until after dark. The last troops were embarked shortly after 1700, and *Bexar* then ordered the remaining LSTs to clear the harbor. *Bexar* (escorted by *Foss*) left shortly thereafter but, unfortunately, grounded near the southern end of the swept channel, and had to wait for high water the next morning to get clear. Shore bombardment by the destroyers began at 1735 while the last LSTs were endeavoring to clear the harbor.[25]

HMCS *Cayuga* with her 4-inch guns and HMAS *Bataan* with larger 4.7's concentrated their fire on oil storage tanks, supply dumps, loaded freight cars, railway lines, and pier installations. *Forrest Royal*, with heavier 5-inch guns, was assigned sturdier, less destructible targets such as a massive brick chimney in the harbor. When repeated hits by 5-inch rounds did not weaken it appreciably, she shifted her efforts to marshalling yards, rolling stock and other smaller targets on which her accurate fire was very effective.[26]

The combined destruction wrought by the three destroyers was vividly described by Captain Brock in his Patrol Report:

> The fires started by the military were sufficient, in the dusk, to indicate very clearly the target areas, and more fires were started by the opening broadsides. As darkness fell, the fires became more and more brilliant, making the whole waterfront as bright as day; great balls of fire rose from the explosions to dissolve into the huge pall of black smoke which was drifting slowly to the southeast. Whenever a fresh oil tank was hit, which was often, sprays of molten glowing metal were radiated in all directions, adding spectacularly to the show. Fires were observed and explosions heard until 0615 the following morning.[27]

The bombardment ceased in early evening, and the three destroyers proceeded downriver to join *Athabaskan* at the anchorage. Three ROK naval vessels—*Daejeon* (JMS-301), *Tongyeong* (JMS-302), and *Daegu* (JMS-303)—which had been patrolling upriver were, with difficulty, persuaded to accompany the destroyers. (Their commanders felt they stood a better chance of engaging the enemy where they were.) These craft patrolled the anchorage that night and checked refugee-filled boats that were still streaming past. The decision to keep the destroyers in the river overnight proved advantageous. Many of the Shipping Control Administration Japan (SCAJAP) tank landing ships had run aground near the anchorage, and others had elected to remain there rather than risk the channel in darkness. (Japan was never formally part of the United Nations' force in Korea. However, operating under the orders of Commander Naval Forces Far East, SCAJAP provided logistic support for the Occupation forces with its fleet of Japanese-crewed ships, most of which were former U.S. Navy Landing Ships Tank.)[28]

The destroyers left the anchorage at first light on 6 December. The grounded LSTs refloated at high water, and those that had anchored for the night were collected, and the long procession of destroyers and LST's, led by *Athabaskan*, set off downriver. *Cayuga* brought up the rear, enjoining the LSTs to greater efforts with a loudhailer. All ships were clear of the channel by 0945, and Brock was able to report, "Mission successfully completed."[29]

AWARDS FOR VALOR

DSO DSC British Empire Medal Legion of Merit Distinguished Service Medal

Three individuals serving aboard the Canadian destroyer HMCS *Cayuga* (DDE218) received awards for valor. Capt. Jeffry Vanstone Brock, RCN, was awarded the Distinguished Service Order (DSO) "for courage, initiative and vigorous leadership during the Chinnampo evacuation while in command of HMCS *Cayuga*." His navigator, Lt. Andrew L. Collier, RCN, received the Distinguished Service Cross (DSC), as previously noted. Additionally, Chief Petty Officer First Class Douglas James Pearson, RCN, was awarded the British Empire Medal per the Canada Gazette, of 14 July 1951, "CPO1 Pearson was the coxswain of HMCS *Cayuga* during her voyage up the estuary known as the Daido-ko to the port of Chinnampo." (As *Cayuga*'s coxswain, Pearson would have been on the wheel at critical times like 'Action Stations' or 'Special Sea Dutymen.')[30]

Capt. Samuel G. Kelly, USN, was awarded the Legion of Merit for exceptionally meritorious conduct in the performance of outstanding services to the Government of the United States as Commander Transport Squadron ONE, from 4 to 26 December 1950. Rear Adm. Lyman Augustus Thackrey, USN, received the Distinguished Service Medal "for exceptionally meritorious and distinguished service in a position of great responsibility…as Commander, Amphibious Group THREE during operations against enemy forces in the Korean area from 12 September 1950 to 1 January 1951."[31]

9

Departure/Arrival of Ships in Theater

> *The French ship would receive many visits of many US Officers of high rank, coming aboard often with impressive staff. La Grandière had become quickly famous in all the Yellow Sea because of the extraordinary "wine cellar" of her Captain CF Urbain CABANIÉ FN. Respectfully towards all these high ranked RN, USN or USMC Officers and their numerous staff, I would [like] to say that we have aboard the La Grandière a high trained team to secure perfectly well [assist] our visitors all along their perilous going down on our accommodation ladder to their motor launches [following festivities on board] weather conditions becoming often suddenly bad in these seas...*
>
> *[HMS] Mounts Bay in company with [FS] La Grandière had previously been engaged on convoy escort duties and when in harbour, cemented this working relationship by the exchange of rum, wine, gauloises [cigarettes] and hospitality. Red wine mixed with rum produced a 'tot' of 200% octane rating, more suited to aircraft fuel than social drinking but even so very much appreciated...*
>
> —Léon C. Rochotte, former *La Grandière* crewmember, describing in separate accounts, attractions of the French frigate to UN Naval Force personnel, both officers and enlisted men, alike.[1]

On 2 October 1950, Vice Adm. Arthur D. Struble, aboard his flagship, the cruiser USS *Rochester* (CA-124) at Inchon, ordered all Seventh Fleet minesweepers under way for Wonsan as soon as possible. Having finished their participation in the Inchon invasion, the French frigate *La Grandière*, as well as British frigates *Whitesand Bay* and *Mounts Bay*, New Zealand frigates *Tutira* and *Pukaki*, and two ROK minesweepers were assigned to Task Group 95.6 (minesweeping group) under the command of Capt. Richard T. Spofford USN. As the minesweepers began sweeping into Wonsan (sustaining loss of four of their craft), the frigates were assigned support roles, which included patrol off northeast Korea and working with ROKN vessels, searching inlets and bays for minelaying junks and sampans.[2]

Photo 9-1

At a conference aboard USS *Dixie* (AD-14), on 5 January 1951, Vice Adm. Arthur D. Struble, USN, commander, Seventh Fleet, (center) points out a possible target area for naval gunfire support missions to Vice Adm. William G. Andrewes, RN, (left) and Rear Adm. Allan E. Smith, USN. Andrewes and Smith were the senior officers of the UN Blockading and Escort Force (Task Force 95).
National Archives photograph 80-G-424769

Following the opening of Wonsan, the aforementioned frigates (with the exception of *Mounts Bay*) left Inchon, on 17 October, escorting five troopships carrying a brigade of Marines. The convoy arrived at the entry channel to Wonsan, on the 21st, and was held there until the 25th when given the go-ahead to proceed. The ships passed through the channel at low speeds, slowing even more and occasionally stopping, upon sighting floating mines.[3]

After landing the Marines at Wonsan without any enemy opposition, as the port had already been secured by ROK ground forces, the frigates set a course for Sasebo, joining *Mounts Bay* at sea. When they arrived, on 27 October, it was cold and snowing. *La Grandière* remained at Sasebo until receiving orders, 15 November, to transport naval personnel to Chinnampo. On the 19th, she anchored off that port to land her passengers.[4]

The night of 21 November, *La Grandière* received orders from French Vice Adm. Philippe Auboyneau, commander in chief, Far East

Naval Forces (Forces Maritimes en Extrême-Orient, FMEO), recalling her to Indochina, as a result of the Cao Bang disaster. She departed the theater, on 25 November 1950, ending the participation of the French Navy in the Korean War.[5]

Photo 9-2

Capt. Willard A. Saunders, USN, commander, Destroyer Squadron 30, Vice Adm. Philippe Auboyneau, commander in chief, French Naval Forces Far East, and Col. Erma A. Wright, USMC, US Naval Attaché Saigon, Vietnam.
Naval History and Heritage Command photograph #NH 74118

BATTLE OF CAO BANG

The battle of Cao Bang, in October 1950, was linked to creation of the People's Republic of China (PRC), in October 1949, and associated actions that followed its inception. In January 1950, the PRC officially recognized the Democratic Republic of Vietnam (DRV), and the Soviet Union followed suit. From May 1950, Chinese military aid and advisors began to cross into DRV territory, internationalizing the Franco-Vietnamese war and changing the military nature of the conflict.⁶

Chinese and Vietnamese strategists knew that Gen. Georges Revers, chief of the Combined Chiefs of Staff for the French Army, had visited Indochina, in mid-1949, and had concluded that the French should not try to hold the northern border against a potential Chinese attack and should accordingly withdraw troops from Cao Bang and Lang Son. Rather than directly attacking French border posts in these two areas, Gen. Vo Nguyen Giap, who led the Viet Minh, decided to wait until the French began withdrawing along vulnerable narrow roads, favorable to a Vietnamese attack.⁷

When the French began to withdraw their troops, on 3 October 1950, in two separate columns, Vietnamese troops attacked the main French column as it withdrew to the east along Route Coloniale 4. Giap employed artillery and two regiments of the 308th division against the retreating troops. French efforts to relieve the column from Dong Khe and That Khe proved futile owing to the inhospitable terrain and the intensity of the Vietnamese attacks. Within two weeks, the DRV prevailed in its first major battle victory over the French.⁸

The French loss at Cao Bang coincided with the entry of Chinese troops into the Korean War, and the French and the Americans began to view the wars in Indochina and Korea as part of the same front designed to hold the line against Chinese communist expansion.⁹

ROYAL THAI NAVY AND COLOMBIAN NAVY SHIPS

Photo 9-3

Thai frigate HTMS *Bangpakong* at Bangkok in 1967, with the sub-chaser *Tongpliu* in the background. *Bangpakong* was the former HMS *Burnet*, and *Tongpliu* the ex-USS *PC-616*. Naval History and Heritage Command photograph #NH 96081

Photo 9-4

Thai transport *Sichang* at Singapore in 1978.
Naval History and Heritage Command photograph #NH 96082

The departure of French frigate *La Grandière* from Korean waters would have dropped the number of nations participating in the UN Naval Force from eight to seven, save for the arrival of the Royal Thai Navy frigates HTMS *Bangpakong* and *Prasae*, on 7 November. Six weeks earlier, on 22 October, a Royal Thai Naval Unit—consisting of the two frigates, the transport HTMS *Sichang*, and a chartered commercial vessel (the *Hertmaersk* from Denmark)—left Bangkok. Aboard the ships was the first contingent of the 1st Battalion, 21st Royal Thai Regiment. The ships arrived at Pusan, on 7 November, where the ground troops disembarked to the UN Army camp at Daegu, located northwest of the seaport. Upon completion of the movement of the soldiers and supplies ashore, the unit sailed for Sasebo, on the 9th.[10]

Following their arrival in Japan, the frigates were assigned to escort duties, which lasted until the end of December 1950. HTMS *Prasae* and *Bangpakong* arrived in the Sea of Japan, on 3 January 1951, for operations with UN forces off the east coast of Korea. As previously discussed in Chapter 4, during a snowstorm, *Prasae* grounded off the coast of Sokcho, on 7 January, resulting in the deaths of two crewmembers and injuries to twenty-three others. Unsalvageable, the frigate was sunk to prevent its capture by enemy forces.[11]

Photo 9-5

Thai frigate HTMS *Prasae*, formerly HMS *Betony* (K274), stranded in the surf off the east coast of Korea. A helicopter and several U.S. Navy ships, including the destroyer minesweeper USS *Endicott* (DMS-35), are offshore covering salvage operations. National Archives photograph #80-G-432568

All, or nearly all of the service of three other Thai frigates that would serve in Korea, and of three from Colombia yet to arrive, would be on the east coast of Korea. The Colombian frigate ARC *Almirante Padilla* did have an initial short stint of duty on the west coast, following her arrival at Sasebo, on 5 May 1951. Assignment of these six frigates to USN task force commanders made perfect sense. The United States and Colombia were neighbors in the Americas, and the U.S. and Thai navies had experience operating together in the Far East. Also, Thailand's two newest ships and all three of Colombia's, were ex-USN *Tacoma*-class frigates, with commonality to the U.S. Navy. By this same logic, it made perfect sense for the ships of the Commonwealth navies—RN, RAN, RNZN, and RCN—and of the Netherlands, to work together on the west coast.

Ships of the US and ROK navies were employed on both coasts, while the cruisers, destroyers, and frigates of the European, and Commonwealth navies mostly engaged in blockade and patrol, shore bombardment, and the screening of RN and RAN carriers off the west coast. As might be expected, ROK vessels off both coasts, worked inshore waters, with which they were well familiar.

An ongoing requirement for patrol, shore bombardment, and mine clearance on the east coast, demanded the attention of most of the American warships in theater. Those dispatched to the west coast were mostly for minesweeping, and screening of USN light and escort carriers. Shore bombardment and patrol were carried out as necessary to complement the other UN Forces. The degree of support and cooperation of UN Forces is evidenced in the photograph, below, by frigates from four navies alongside the repair ship USS *Jason*, in the Han Estuary, west coast of Korea, on 16 January 1952.

Photo 9-6

Left to right: HMAS *Murchison*, ARC *Almirante Padilla*, USS *Gloucester*, and ROK *Taedong*. U.S. Navy photograph courtesy of NavSource

10

Abandonment of Inchon, NW Islands Home to Guerillas

On 30 November 1950, Vice Admiral Joy sent a request to Adm. Eric J. P. Brind, RN (commander in chief, Far East Fleet), that Rear Admiral Andrewes return from Hong Kong, and that all HM (His Majesty's) ships that had not been stood down be made immediately available. On 1 December in Hong Kong, Andrewes hoisted his flag in the light carrier HMS *Theseus* and with the destroyer HMS *Constance*, sailed for Sasebo. On arrival there three days later, he resumed the duties of commander, Task Group 95.1 from Captain Lloyd-Davies. Responsible for all west coast areas, an immediate requirement for Andrewes task group was the support of evacuation of units of the Eighth Army from Chinnampo (detailed in Chapter 8).[1]

Throughout 5 December, the evacuation proceeded smoothly and, by early evening, 5,900 ROK troops and 1,800 U.S. Army and Navy port personnel had been evacuated. HMS *Theseus* provided air cover for the Commonwealth, and U.S. ships engaged in the evacuation, and the next day, they left the Chinnampo area bound for Inchon.[2]

Upon his arrival at Inchon, on 5 December, Andrewes specified the duties of Task Group 95.1 as follows:

- West Coast blockade
- Anti-aircraft defense of Inchon and naval gunfire support
- Air cover over the task group and armed reconnaissance north of the bomb-line (designated area in which Allied ships carried out shore bombardment)[3]

Andrewes believed that the Chinese would not restrict their efforts to ground combat, and considered it highly probable that they would strike at naval forces by aircraft and submarine. His concern stemmed, in part, from an incident in early November, when MiG-15 fighters had attacked Allied bombers in the Yalu area of the west coast.[4]

As a further precaution against air attack, Andrewes reorganized his task group into four Task Elements:

- TE 95.11: To operate west of Inchon – light carrier HMS *Theseus* (CTG 95.1 embarked), with the destroyers HMS *Cossack*, *Constance*, and *Concord* screening her
- TE 95.12: Surface Blockade and Patrol Element – cruiser HMS *Ceylon* (CTE 95.12), and destroyers HMCS *Cayuga*, *Sioux*, and HMAS *Bataan*.
- TE 95.13: At Inchon; available for various duties as required – frigates HMS *Cardigan Bay* (CTE 95.13), *Morecambe Bay*, and HMNZS *Tutira* and *Rotoiti*
- TE 95.14 At Inchon, primarily for anti-air defense – cruiser HMS *Kenya* (CTE 95.14), and destroyers HMAS *Warramunga*, HNMS *Evertsen*, and HMCS *Athabaskan*[5]

For the remainder of the month of December, the blockade patrol was maintained by TE 95.12 between Inchon and the Yalu Gulf, with task element ships periodically exchanging duties with those of the Inchon Gunfire Support Element (TE 95.13). Operations in the Inchon area involved: covering the seaward flank of the Eighth Army by air and surface blockade; conducting air strikes over northwest Korea; and carrying out armed air reconnaissance.[6]

During this period, the blockading force became aware of the presence of North Korean guerillas, occupying northwest islands, who had fled the mainland, and who wanted to fight the Communists. First contact with the guerillas had been made earlier by ground forces.

FIRST CONTACT MADE WITH UN GROUND FORCES

As UN troops forged northward in the autumn of 1950, armed guerillas from North Korea's Hwanghae province made contact with the advancing forces, protected their lines of communication, and helped in harrying the retreating North Koreans. This interaction marked the start of the involvement of the U.S. Eighth Army with unconventional warfare that would take place along the rugged coastline and offshore islands of the Hwanghae province of North Korea. The population of Hwanghae was culturally and politically aligned with Seoul and, when the Chinese attacked southward, the guerillas were forced to flee, along with thousands of refugees and irregular forces from other parts of North Korea.[7]

Some of the Hwanghae guerillas remained in the province, and carried on resistance from the mountains; some established strongholds along the Hwanghae coast; and others took refuge on offshore islands. Many fled to the island of Cho-do, southwest of Chinnampo, by way of the village of Wolsa-ri. Others escaped southwest through the town of Changyon into the small Changsan-got peninsula, and then south to the Paengyong-do, Taechong-do, and Sochong-do islands. (See Map 5-3.)[8]

Map 10-1

Area of North Korea surrounding the Hwanghae peninsula
Source: https://campsabrekorea.com/maps-of-korea-1950s--present.html

ASSISTANCE FROM ROK NAVY

Throughout December, a flood of escapees fled to islands off northwest Korea in sampans and junks, with some assistance from the Republic of Korea Navy. Badly needed help dramatically increased in early January 1951, when ROKN ships (and a merchant marine LST) evacuated refugees from the southwest coast of Hwanghae, as partisans delayed advancing Chinese and North Korean forces. On 19 January, the ROK Navy evacuated some 13,000 North Korean refugees and guerilla fighters from the northwest Korean coast—supporting the operation with heavy shore bombardment and an amphibious landing south of Chinnampo. Among the masses taken off, was the Pyeongyang Partisan Regiment. Based in northwestern Hwanghae province, it had fought its way to the coast, and was withdrawn to Cho-do Island (while North Korean Army forces cut off and attacked a second group of irregulars from the same area of the province).[9]

TASK GROUP 95.1 AWARENESS OF GUERILLAS

Task Group 95.1's relationship with the anti-Communist guerillas began in mid-December 1950, but in a very indirect way. Intelligence obtained from ROK naval craft, which was collected from guerilla groups, was the only communications connection between the blockade ships and the North Korean freedom fighters.[10]

At the time, task element commanders engaged in blockade duties, were not able to devote much attention to the anti-Communist activities emerging within their operational areas, owing to their involvement in supporting ground forces retreating south of the Han River. Then, TG 95.1 viewed the large number of guerillas as only uncontrolled North Korean irregulars with potential as a useful intelligence source.[11]

During his patrol as commander, TE 95.12, from 4-18 December, Captain Lloyd-Davies, commanding officer of the cruiser HMS *Ceylon*, acquired some valuable and up-to-date intelligence about Communist activities. This was supplied by Captain Choi Hyoyong of the patrol frigate ROKN *Dumon* (PF-61), which was attempting the rescue of refugees from the mainland. Lloyd-Davies was impressed with the value of the intelligence, which had been previously unavailable.[12]

Photo 10-1

Capt. Cromwell F. J. Lloyd-Davies (2nd from right) in September 1945, aboard the heavy cruiser HMS *Sussex* at Singapore, for the Japanese surrender of the British Crown Colony. British representatives are: Lt. Frank P. Donachie, RNVR (Flag Lieutenant); Maj. Gen. Eric C. R. Mansergh, BA; Brigadier Norman D. Wingrove, IA; Lt. Gen. Alexander F. P. Christison, BA; Rear Adm. Cedric S. Holland, RN; Captain Lloyd-Davies (Chief of Staff); and Maj. Gen. Herbert R. Hone, BA (Civil Affairs Officer). Australian War Memorial photograph 041572

ROK Navy access to new sources of intelligence was welcome news for the crews of blockade ships, which had wanted to develop their own intelligence collecting system. Lloyd-Davies arranged with the

ROK Navy to make procuring intelligence a major part of its activities. Choi Hyoyong agreed to assign one of his officers to serve as intelligence officer, and arranged to send North Korean guerillas onto the mainland to collect information. If any info of particular value was gathered, he would signal a request for a rendezvous to pass it to Lloyd-Davies.[13]

Photo 10-2

Pointing out the opened breech of a gun aboard HMS *Ceylon*, is Marine F. N. Barker (far right) to his brother-in-law Lt P. Dooks (Duke of Wellington Regiment), who had a chance to visit the cruiser when *Ceylon* was in drydock at Kure, Japan, 10 July 1951. Australian War Memorial photograph DUKJ4557

This relationship continued during the subsequent blockade patrol. Captain Patrick W. Brock of the cruiser HMS *Kenya*, which had relieved HMS *Ceylon*, frequently met Choi Hyoyong on Techong-do and received information about the Haeju and Chinnampo areas.[14]

SECOND CHINESE OFFENSIVE

The great desire for information about enemy activities and targets ashore was spurred by a second Chinese offensive, which began on 7 December 1950 and lasted through 25 January 1951. As large numbers of Chinese forces continued to advance south, in early December, the UN Command contemplated resistance by Eighth Army troops in the area of Seoul, with subsequent retirement to Pusan. The situation was dire, as witnessed by the assignment of Navy underwater demolition teams to survey beaches in South Korea, Tsushima Island, and western Japan in preparation for a possible emergency withdrawal.[15]

At Inchon, Rear Adm. Lyman A. Thackrey was scouting the nearby Tokchok islands southwest of Inchon, as a possible refuge, in the event of an emergency evacuation and redeployment. On 6 December, with the evacuation of Kimpo Airfield looming, he asked Vice Admiral Struble for carrier air support. The following day (two days before Tenth Corps, on the east coast, received orders on the 9th to redeploy south from Hungnam), Thackrey was instructed to begin removing army supplies from Inchon. Soon after Eighth Army, as it retired southward, would require naval gunfire support along the entire western coast of South Korea.[16]

Amongst these gloomy circumstances, Lt. Gen. Walton Walker was killed in a road accident, requiring Lt. Gen. Matthew B. Ridgway to be sent from Washington, D.C., to take over Eighth Army. Such duty must not have been a promising proposition for him. MacArthur's early dispatches had produced an atmosphere of depression in the halls of power, and there had been setbacks and delays in actions intended to provide relief. UN efforts to seek a cease-fire agreement, had ended with Chinese rejection of the terms proposed. President Harry S. Truman had declared a state of national emergency, and efforts to increase the nation's armed strength were redoubled. Meanwhile, the available reserves within the continental United States, continued to be one Army and one Marine division.[17]

On 26 December, as Ridgway arrived in Korea, Tenth Corps was integrated into Eighth Army. At Pusan, the last of the forces evacuated from Hungnam were going ashore. Some troops from Hungnam were sent to the west coast, where an expected enemy attack on Seoul was awaited. In related actions, the escort carriers USS *Sicily* and USS *Badoeng Strait* relieved the light carrier HMS *Theseus* in the Yellow Sea operating area, on 27 December, and began to fly missions in support of Eighth Army. Two days later, commander, Cruiser Division One, Rear Adm. Roscoe H. Hillenkoetter, USN, arrived at Inchon with the heavy cruiser USS *Rochester* (CA-124), to join HMS *Ceylon* and the Australian destroyers HMAS *Warramunga* and *Bataan* in support of land forces on the Kumpo peninsula, located west of Seoul.[18]

With the arrival of the New Year, three Chinese armies pushed down the northern approaches to the capital city of Seoul. In the center of Korea, another heavy thrust was delivered against Eighth Army north of Wonju, a city 87 miles east of Seoul. On 4 January 1951, Seoul was abandoned, the Han River bridges were blown, and the army started south again. At Inchon, all ships were put on one-hour notice and, on orders from Vice Admiral Joy, destruction of the port was begun.[19]

Photo 10-3

Heavy cruiser USS *Rochester* (CA-124), circa December 1952-March 1953. Naval History and Heritage Command photograph #NH 96884

INCHON ABANDONED, PORT OF TAECHON OPENED

On 5 January, the Inchon port facilities were blown and, as the Chinese entered the town, Thackrey sortied his shipping. In the last five days of an effort that had begun the previous month, a few hundred more vehicles, a few thousand more tons of supplies, and another 37,000 military personnel were taken out.[20]

With Inchon gone, and only overloaded Pusan remaining, to which the larger ships were sailing, a problem of supporting the western flank without overwhelming the Pusan port organization, and the rail and road systems, arose. A major port further north was needed, and this requirement was met by opening a seaport where none had existed before. Twenty-five miles north of Kunsan, the town of Taechon lay at the head of a bay and, from it, a road and single-track railroad ran northeast, joining the main line at Chonan, behind the newly formed Eighth Army front.[21]

Prior to the landings at Inchon in September, MacArthur's need for reconnaissance of potential beaches, for a major amphibious assault behind North Korean lines, had brought the UDTs (underwater demolition teams) from USS *Horace A. Bass* to seek a landing place at Taechon. In December, as a precautionary measure, Thackrey had swept a major anchorage area there, and, at the time of the evacuation, following a check-sweep by *Carmick* and *Swallow*, the tank landing ships from Inchon were beached and the servicemen and stores aboard them unloaded.[22]

Throughout the retirement of Eighth Army, UN naval forces did what they could to help stem the Chinese tide. At Inchon, *Rochester*, *Kenya*, and *Ceylon* supported the withdrawal of troops across the Han and evacuation of the port, and shelled nearby Kimpo Airfield. From the Yellow Sea, US Marine fighter pilots aboard *Sicily* and *Badoeng Strait* provided protective patrols overhead, struck the advancing enemy, and destroyed quantities of abandoned supplies at Kimpo. On 1 January, Eighth Army's need for increased support precipitated an increased carrier strength for the area. Two days later, the light carrier USS *Bataan* (CVL-29) arrived to join the west coast group.[23]

Photo 10-4

USS *Bataan* (CVL-29) with F4U-4B Corsair fighter-bombers of VMF-314 on deck. National Archives photograph #80-G-633888

Just when prospects of UN success in Korea appeared dimmest, and evacuation from the peninsula increasingly probable, signs began to appear that the Chinese had overreached in their offensive. Planes from carriers off the west coast reported ROK flags flying in coastal villages, while the governor of Hwanghae province asked for ammunition. On 13 January 1951, Rear Adm. Allen E. Smith (CTF 95) recommended that Vice Adm. C. Turner Joy (ComNavFE), arm the estimated 10,000 patriotic volunteers in this area.[24]

General Ridgway, the new commander of Eighth Army, instead of looking over his shoulder toward Pusan, was directing his gaze northward. On 16 January, a reconnaissance in force had penetrated as far north as Suwon, in northwestern South Korea. Nine days later, the northward movement of Eighth Army began, on 25 January, against only slight resistance. Ten days later, the Chinese commander decided to retire northward of the 38th parallel.[25]

TASK FORCE 95.1 LIAISON WITH GUERILLAS

In January, although the primary naval requirement on the west coast was support of Eighth Army, blockade commanders continued to try to monitor guerilla activities as a means of obtaining intelligence. After returning to patrol duties from gunfire support for ground forces at Inchon, HMS *Ceylon*'s first action as flagship of Task Element 95.12, was to visit Techong-do and contact the senior ROKN officer there to gather information about enemy activities and targets ashore.[26]

On 19 January, active utilization of guerillas for blockading force purposes was attempted by Sir Vice Admiral Andrewes (earlier promoted and knighted on 1 December). Andrewes had received a message from CTF 95, stating that the enemy was operating a night-time train between Haeju and Ongjin and instructing him to 'catch that train'. In response, Andrewes decided to use the guerillas based on the mainland, and sent instructions to HMS *Kenya* to inquire whether the guerillas could destroy the train.[27]

This spurred Capt. Patrick W. Brock, RN, of HMS *Kenya* meeting with the commanding officer of the ROKN patrol craft *Chiri San* (PC-704) to propose the operation. The South Korean naval officer told him that he was not in communication with any guerillas in the south of Hwanghae province, and doubted whether there were any with proficiency in demolition. Accordingly, Brock reported to Andrewes that the guerillas were unable to carry out the proposed operation.[28]

UNITED STATES EIGHTH ARMY INVOLVEMENT

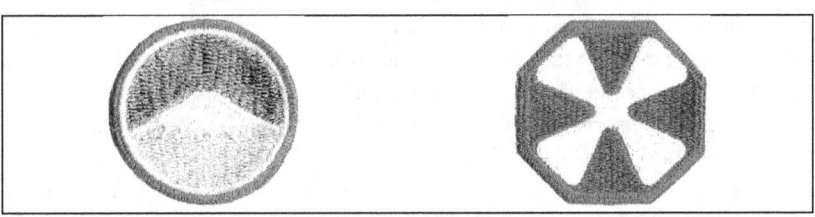

Far East Command patch (left) and on right, U.S. Eighth Army shoulder sleeve insignia

Although General MacArthur's Far East Command (FEC) and U.S. Eighth Army leadership knew about the exodus of civilian refugees from North Korea, they were unaware, until 8 January, that guerillas were among them. They learned about this when Task Group 95.7 (ROK naval forces) reported that approximately 1,000 volunteers with

Japanese arms were operating in Hwanghae province and asked if Eighth Army had any ammunition that could be made available.[29]

The plea for help found an advocate in Eighth Army staff officer Col. John McGee. McGee had conducted a successful guerilla campaign in the Philippines during World War II and had earlier tried, in July 1950, to persuade Eighth Army to arm and equip a behind-the-lines guerilla force in Korea. His earlier proposal was rejected, but now found support for establishing an organization to assist, organize, and direct the Hwanghae irregulars (Army term for guerillas). On 15 January 1951, the Eighth Army G3 Miscellaneous Division, which had been coordinating special operations in Korea, established an Attrition (Partisan) Section to direct the efforts of the North Korean irregulars.[30]

On 15 February, Eighth Army established a base of operations for the partisans on the west coast island of Paengyong-do. Located in a remote location (the westernmost point in South Korea), the island was already in use by the CIA, the South Korean government, and the U.S. Air Force for their own clandestine operations. Eighth Army also formed two supporting organizations: Baker Section, with headquarters and training camp near Pusan, to conduct airborne training and insertions; and Task Force Redwing, an American-led company of ROK Marines that carried out intelligence, sabotage, and commando operations along the coasts and among the islands off North Korea. Originally code-named William Able, the Paengyong-do base was renamed Leopard, in March 1951, and the partisan units adopted the nickname "Donkey" with a unique number to identity each one.[31]

The William Able and (later) Leopard operations were planned, directed, and supplied by Eighth Army staff. In carrying them out, the guerillas reached the mainland shore from islands off the coast utilizing motorized junks, and smaller craft, including sailing junks and little boats powered by a single oar sculled over the stern. Some of the larger craft were fitted with an additional powerful engine, hidden radio antennas, and concealed recoilless rifles.[32]

Supported periodically by ships of the Royal Navy, the irregulars conducted raids against Chinese and North Korean forces and carried out intelligence collection, aircrew rescue, sabotage, and assassination missions through war's end. The partisan operations were tactically effective in producing enemy casualties, but were essentially limited to Hwanghae province.[33]

11

Blockading Force/ Guerilla Cooperation

Photo 11-1

Destroyers of British Commonwealth navies moored to a buoy at a southern Japanese port (presumably Sasebo), circa May 1951, following an extended period of operations as members of the UN Blockading and Escort Force. From left: HMAS *Warramunga*, HMCS *Nootka*, and HMS *Cockade*.
Naval History and Heritage Command photograph #NH 97046

From late January 1951, the ground war was changing as UN forces counterattacked, and the recapture of Inchon on 10 February followed steady advancement northward. Throughout this period, naval forces were heavily involved in supporting left flank ground operations. The British cruiser HMS *Ceylon* and Canadian destroyers HMCS *Nootka* and *Cayuga* were employed for naval gunfire support, and interdiction and harassment fire. The cruiser HMS *Belfast* subsequently took over command responsibility from *Ceylon*. Task Group 95.1's involvement in this role continued until UN ground forces recaptured Inchon and then, as the immediate requirement for naval gunfire support on the left flank was relieved, and ground forces continued to push northward, the task group was able to concentrate more on the west coast blockade.[1]

It is important to note that although most reference to ground forces in this book are to the U.S. Eighth Army, many UN countries contributed troops to the conflict in Korea. For example, the newly formed 2nd Battalion, Princess Patricia's Canadian Light Infantry, under Lt. Col. James Riley Stone, arrived at Pusan, on 18 December 1950. By 23 February, their training completed, the "Pats" were "in the line" with their comrades of the 27th British Commonwealth Brigade. Together with their brothers of the Australian 3rd Royal Regiment, they earned U.S. Presidential Unit Citations in the Battle of Kapyong, in April 1951, where they played key roles in stopping Chinese advances in mountains near the 38th parallel.[2]

Photo 11-2

Brigadier R King CBE DSO inspects the Princess Patricia's Canadian Light Infantry at Kure, Japan, in March 1951. At far left is Captain E Sharte, the officer commanding the Canadian detachment.
Australian War Memorial photograph DUKJ4167

Later in the month following the abandonment of Inchon, on 5 January, cruisers and destroyers were sent there to bombard enemy installations in the area. This action was undertaken in the hope that it might mislead the enemy into thinking that another amphibious assault was planned, and cause them to divert troops there away from the main assault. At Inchon, HMCS *Cayuga* and *Nootka* came under enemy fire for the first time. As they were leaving Inchon harbor, on 25 January, enemy guns on Wolmi-do opened fire on them. Both ships immediately reversed course and closed to engage. After their 4-inch guns silenced the shore batteries, they then, for good measure, steamed in to Bofors range to spray the area with the short-range weapons, before departing to resume patrol duties.[3]

Photo 11-3

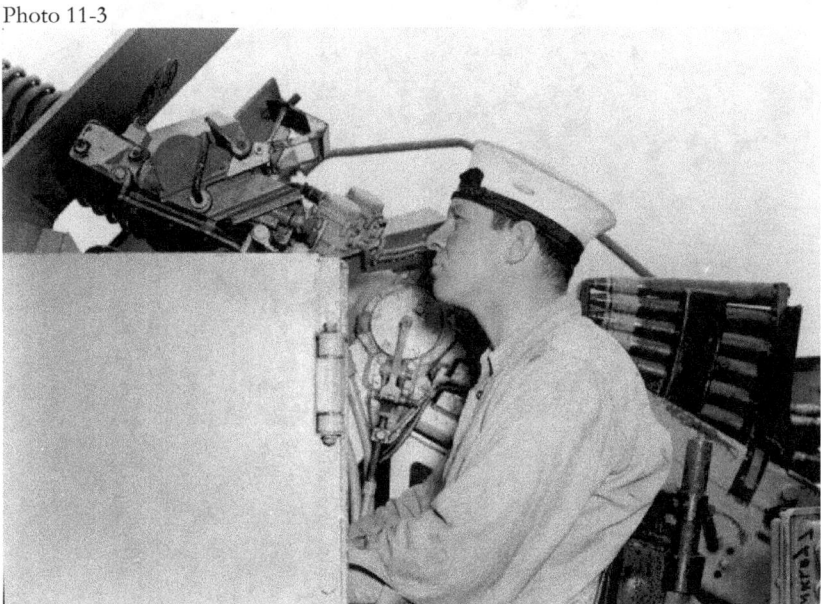

Aboard HMCS *Nootka*, a seaman mans a 40mm Bofors gun during non-combat conditions (no wearing of battledress), with "ready use" ammunition to his right. Australian War Memorial photograph DUKJ4080

The following day, 26 January, the heavy cruiser USS *St. Paul* (CA-7) was also fired upon by shore batteries at Inchon.[4]

HMCS *Cayuga* returned to Inchon, on 27 January, to join the UN bombardment force in the harbor. While General Ridgway was making good progress in his first offensive, the ships were under orders to harass the enemy, disrupt his lines of communications and, when possible, provide direct gun-fire support for UN troops advancing up the west coast of the Korean peninsula.[5]

On the afternoon of 30 January, *Cayuga* came under enemy fire for the second time in less than a week, when shore batteries (believed to consist of about six 75mm guns) opened on her. Several rounds landed within 200 yards of the ship, but the combined fire power of two cruisers and two destroyers were too much for the artillerymen, and their guns soon fell silent. *Cayuga* remained at Inchon, until 3 February 1951.[6]

Photo 11-4

Heavy cruiser USS *St. Paul* (CA-73) under way, location and date unknown. Naval History and Heritage Command photograph #L45-248.07.01

AMPHIBIOUS DEMONSTRATION CANCELLED, AND UN FORCES RESUME USE OF THE PORT OF INCHON

Following the landings at Inchon, in September 1950, the enemy was well aware of how decisive, and deadly an amphibious assault could be. To take advantage of this sensitivity and, in an effort to cause diversion of troops from the front to landing areas, the U.S. Navy would conduct many amphibious demonstrations (feints) during the war, beginning with one on the east coast near Kansong, on 30-31 January 1951.[7]

On 8 February, in preparation for a fake landing at Inchon two days later, the battleship USS *Missouri* (BB-63) "opened the show" with shore bombardment. Further operations became unnecessary owing to the rapid advance of UN ground forces, which forced enemy evacuation of the Inchon area. It is likely that the prospect of a second invasion at Inchon, suggested by barrages from the battleship, may have made the enemy's departure of the area more urgent and rapid. On 10 February, all opposition on that front disappeared.[8]

Photo 11-5

Crewmen handle 16-inch projectiles aboard the battleship USS *Missouri* (BB-63). Naval History and Heritage Command photograph #NH 96784

Seven weeks later, commander, Amphibious Group and 2nd Engineer Special Brigade commenced operations on 28 March for the reopening of Inchon as a resupply port.[9]

INITIAL ORGANIZATION OF GUERILLA FORCES

From mid-January 1951, the ROK Navy undertook efforts to organize and employ the guerillas that had fled to Paengyong-do. This island lay approximately 15 miles offshore, southwest of the Hwanghae peninsula, and two miles south of the 38th parallel. It was large enough (almost 18 square miles) to support large numbers of troops, and offered other attractions to UN forces. These included prominent hills where radar and radio antennas could be installed, and a long strip of hard-packed sand, and sufficient depth of water in the island approaches, for a tank landing ship to beach, at mid-to-high tide.[10]

At that time, a company of ROK Marines and 30 ROK Navy personnel under the command of Lt. Col. Lee Huijeong, were sent to Paengyong-do. Huijeong's first duty was island defense, but he also had orders to organize the guerillas into raiding and intelligence gathering as well as the required garrison parties. To do so, the colonel and his staff selected several hundred loyal anti-Communist youths, and provided them basic military and communication trainings. Needed supplies such as rice, ammunition, and weapons were also furnished. Following instruction, the trainees were split up into several 30-50 member units. Some units were assigned to garrison force duties on the island, while others continued intelligence gathering and guerilla activities on the mainland.[11]

Simultaneously, the ROK Navy tried to increase the number of guerilla groups with which it shared close communications. By supplying gunfire support and ammunition to guerillas in the Cho-do–Sok-to area, contact was initiated with partisans further north. For example, as a consequence of close support by the patrol craft *Daegu* (JMS-303) and *Taebaeksan* (JMS-304), regular contact was made with the largest guerilla group from Mount Kuwol in South Hwanghae.[12]

BLOCKADING FORCE/GUERILLA RELATIONSHIPS

With Task Group 95.1 then able to focus more on its blockade mission, the ROK Navy's success in organizing the guerillas on Paengyong-do (which were admittedly only a small portion of those on the west coast), sparked the interest of the blockade commanders. A relationship between Royal Navy (and other Commonwealth) officers and the guerillas developed rapidly, beginning during the patrol period of Captain Lloyd-Davies as commander, Task Element 95.12 (Surface Blockade and Patrol Element), from 13-24 February 1951. He had been interested in the activities of the guerillas, since December 1950, through intelligence gained from the ROKN, and realized the advantages that might result from raids by the guerillas, supported by TE 95.12 and ROK patrol craft.[13]

The whole area from Inchon to Chinnampo seemed to be lightly held by North Korean troops, employed in a police-like role rather than a defensive one. Moreover, there were many beaches and places to land raiders, with very little to prevent them from walking ashore unopposed. With this in mind, Lloyd-Davies held several meetings with the commanding officers of the ROKN patrol craft *Chiri San* (PC-704) and patrol frigate *Apnokkang* (PF-62), and some of the ROKN officers in command of the guerillas. The result was the planning of a guerilla raid, as detailed in Lloyd-Davies' Report of Proceedings:

> It was agreed that a raid should be made on the village of Kujin-ni (XC 6240), where 150 North Koreans had been reported with one or two field guns, which we intended to capture. If suitable, a follow up attack would be made on Hongga-ri (XC 6523), where there was a further batch of the enemy. The raid was to be conducted in three phases; the first was to land at night some guerillas with Type 68 sets [radios] lent from *Ceylon*, for bombardment spotting; the second was the landing of 200 guerillas in two parties one either side of the town at dawn, followed immediately by the attack, supported by *Ceylon*, *Consort* and *Amethyst*. The whole raid was to last about 3 hours. CTE 95.11 had agreed to support the raid and the necessary planes were laid on.[14]

Unfortunately, an unexpected, strong south-westerly gale resulted in the planned raid being called off. However, the plan developed a model for subsequent Blockading Force efforts to support guerillas, involving naval gunfire and air strikes.[15]

Captain Lloyd-Davies's endeavors to develop active cooperation with the guerillas were continued by subsequent TE 95.12 commanders. Vice Adm. Andrewes, RN, following his promotion, had continued to serve under Rear Adm. Allen E. Smith, USN. On 19 February 1951, as directed by "top brass" in Washington, DC (following a request by Winston Churchill), Andrewes took over the command of Task Force 95 from Smith. Responsibility for blockading and escort duties on both coasts of Korea, now occupied Andrewes' attention. However, without his direct supervision, the issue of cooperation with guerillas on the west coast had to be handled based on each Blockade Commander's discretion.[16]

As a result of increased intelligence gathering, Capt. Aubrey St. Clair-Ford (commanding officer of the cruiser HMS *Belfast*), who took over the patrol, was able to get much target information. In particular, intelligence acquired by ROK Captain Kim's guerilla group in the Cho-do–Sok-to area was very useful. *Belfast* carried out several direct shore bombardments on enemy beach defensive positions, and also indirect bombardment using air-spotting from the light aircraft carrier USS *Bataan*. *Belfast* also passed information to *Bataan* that might be useful for air strikes, acquired from guerillas, and her planes struck these targets as well.[17]

On 4 March 1951, in a meeting aboard HMS *Belfast* with Captain Kim, St. Clair-Ford learned more about him and his group's previous mainland activities and was impressed by their considerable success. St. Clair-Ford, upon being briefed regarding further raids planned in the next few days, decided to support them as best he could. Because HMS *Belfast* was soon to be relieved by HMS *Kenya*, he told Kim to contact the other cruiser to obtain naval and air support if required.[18]

During HMS *Kenya*'s patrol as CTE 95.12, the first gunfire support for a guerilla raid was carried out. Following an urgent call from a ROKN craft on 8 March, which was covering (providing protection for) a landing on the mainland near Cho-do Island, *Kenya* proceeded to the area and employed her much-larger guns as well.[19]

EIGHTH ARMY HAS OTHER PLANS FOR GUERILLAS

In early March 1951, most of the guerilla groups in northwest Korea were still operating with ad hoc command and control. Although the ROK Navy controlled some of them, it was not possible for its staff to do more than support the several hundred guerillas on Paengyong-do, and to maintain sporadic contact with several other guerilla groups scattered on the islands. The ROKN did not have adequate resources to organize large numbers of irregulars under a unified command and control system. On 15 February, an Eighth Army detachment commenced guerilla reorganization on Paengyong-do, but it was not until mid-March that it was able to establish a command system for large groups.[20]

Colonel John McGee selected Paengyong-do as the headquarters for the Eighth Army's Attrition (Partisan) Section—designated Task Force WILLIAM ABLE—and assigned his executive officer, Maj. William A. Burke, to command it. Burke's small staff consisted of three officers, one staff sergeant, and a private. Eighth Army's plans did not directly coincide with the Blockade Commanders' expectation that the new organization would function as a controller and coordinator of all guerilla activities, which would produce valuable intelligence for the naval blockade operation. The army intended to use the guerillas in support of the UN ground forces' operations, which largely restricted their relationship with the blockade ships. Additionally, during the guerillas' activities to form "cell units" ashore, enemy intelligence was collected as a by-product. However, the need for liaison with friendly naval forces for guerilla support also resulted in a closer relationship.[21]

Immediately after Eighth Army recruited the first guerilla unit, it was given 15 days of demolitions and communications training. During this process, guerilla members suggested the name "Leopard" for the overall unit, and "Donkey" for numbered elements of the sub-units. As a symbol of a stealthy and speedy hunter, "Leopard" reflected their desired way of operating. The origin of "Donkey" is uncertain, but is sometimes said to be a derivation of the Korean term dong-il (leader). Burke approved these suggestions, and Task Force WILLAM ABLE became LEOPARD, and sub-units Donkey 1, 2, 3, and so forth.[22]

EIGHTH ARMY PLANNED OPERATIONS COMMENCE

On 1 March 1951, Task Force Leopard (the guerillas operating from the northwest islands) were instructed to initiate partisan operations as soon as possible to support Eighth Army's operations ashore. The first insertion of partisan teams was code named Operation SHINING MOON. The first of the teams to go in, Donkey 1 (comprised of 38 partisans led by Chang Chae Hwa), landed on the Changsan-got peninsula, the night of 3 March, and moved up into the hills to make contact with locals who were willing to join forces with them.[23]

On the night of 20 March, at the time of a meeting of local Communist party officials, Donkey 1 conducted a raid on a police station killing 27, and capturing weapons and ammunition. After other successful attacks on truck stops and warehouses, the team gained new recruits until it was more than 100 strong. It continued operations on the mainland for four and one-half months. By mid-July, the team claimed to have killed some 500 soldiers, police, and Communist officials; blown up bridges; rescued prisoners; and raided warehouses.[24]

These actions resulted in many partisan casualties. On the night of 23 July, twenty of its survivors waded through the low tide to a friendly-held island ending their operation. Other Donkey teams infiltrated during this time, but Donkey 1 was the most successful.[25]

BLOCKADE FORCE INTERFACE WITH GUERILLAS

On 8 March, commander, Task Force 95 received a message from Eighth Army headquarters: "Request RN Officer your TF [TF 95] visit this HQ for conference on support friendly guerilla activities Western Coastal area." Vice Admiral Andrewes replied the next day that he wanted to hold a conference onboard HMS *Belfast*, on 14 March, and requested the Eighth Army send representatives. Captain St. Clair-Ford held the meeting, which was attended by Colonel McGee and Colonel Thompson of the 1st Corps, the commanding officer of the command ship USS *Eldorado* (commander, Amphibious Group 3), and officers from *Belfast*. The discussion involved details about the guerila groups on the islands off the Hwanghae province, and problems of liaison, communication and intelligence dissemination between the Eighth

Army, TG 95.1, I Corps, and the ROK Navy and Marines. It ended with agreement on the following procedures:

> The blockade fleet will periodically put ashore at Paengyong-do a boat to contact Major Burke and secure intelligence of the Hwanghae Area. If it becomes necessary a liaison officer from the blockade fleet will be placed on Paengyong-do. The possibility of netting the Island Radio with the blockade fleet [use of a common radio circuit] will be considered.[26]

Photo 11-6

Cruiser HMS *Belfast*, photographed from HMAS *Bataan*, in December 1950. The frozen sea ice evidences the brutal conditions which UN sailors, soldiers, airmen, and Marines endured during winters in Korea.
Australian War Memorial photograph 042343

From late March, cooperation among the involved parties accrued improved results in four critical areas: intelligence supply, employment of guerillas for naval operations, naval support of guerilla activities, and coordination of guerilla movements. Additional improvements followed. In early April, Task Element 95.12 (Surface Blockade and Patrol Element) began receiving a large number of target intelligence reports from Leopard Headquarters by signals (radio communications) or hand-carried messages. Between 16 and 30 April, during HMS *Belfast*'s patrol as CTE 95.12, she received more than sixty target information items pertaining to the Hwanghae Province and adjacent waters. These included: intelligence on the enemy's offshore activities, such as minelaying and small boat movements; and mainland activities including troop movements, and the location of Battalion Headquarters and anti-aircraft gun positions.[27]

HMS *Belfast* conducted gunfire bombardments based on this target intelligence and also passed on information of possible interest to CTE 95.11, the light carriers HMS *Theseus* and USS *Bataan*.[28]

Photo 11-7

Ice formed on the forward gun mount and other topside areas of the Australian destroyer HMAS *Warramunga*, in wintry conditions in Korea, circa 1952.
Australian War Memorial photograph 305203

In early April, Rear Adm. Alan K. Scott-Moncrieff, RN, succeeded Vice Admiral Andrewes as commander, Task Group 95.1.

Photo 11-8

Rear Admiral Scott-Moncrieff (left) in the engine room of HMAS *Warramunga*, with Chief Engine Room Artificer A. L. McKinnon (center) and Lt. M. McLachlan, RAN (right), while visiting the Australian destroyer.
Australian War Memorial photograph 304835

The Australian destroyer HMAS *Warramunga* relieved the British frigate HMS *Black Swan*, on 11 April, and took over commander, Task Element 95.12 (*Warramunga*, destroyer HMCS *Nootka*, and frigate HMS *Amethyst*) and Senior Officer West Coast Blockade duties. During the ensuing patrol period, Comdr. Otto H. Becher, DSC RAN, of the *Warramunga* instructed ROK craft under his command to concentrate on collecting up-to-date intelligence from Leopard officers.[29]

The first period of the patrol (11-16 April) consisted of organizing patrolling by ROKN units on the coast and gaining intelligence for HMS *Belfast* during operations in the Yalu Gulf. On the 17th, Becher turned over command of TE 95.12 to the commanding officer of *Belfast* and proceeded to the Haeju area to assume control of Korean minesweepers. Supervision of the sweeping operations continued for two days, during which *Warramunga* joined the British destroyer HMS *Comus* for mine search, patrol, and bombardment duties. *Warramunga*'s patrol ended, on 21 April, with operations off the Inchon approaches.[30]

From May 1951, Leopard headquarters began supplying "After Bombardment reports" to TG 95.1, based on information obtained by Leopard agents on the mainland, about such things as number of enemy casualties, and the numbers of structures and equipment destroyed by air strikes. Target intelligence for naval gunfire was mostly obtained by guerillas based on islands, who were usually ashore for only limited periods. Since these periods did not necessarily coincide with shore bombardments, battle damage assessments of gunfire were not reported frequently. In the case of air strikes, guerillas on the mainland had hiding places around the target areas, and could more easily check, so the results of aerial bombardments were better reported.[31]

ONGOING CHALLENGES

Despite improved coordination between Blockade commanders and Eighth Army staff, Task Element 95.12 experienced difficulties regarding unnotified activities of craft operated by Leopard guerillas—mainly a result of Leopard Headquarters' lack of control over guerilla units. Because American army officers were not allowed to accompany guerilla activities on the mainland, until early May 1952, their role in the guerilla organization was mainly one of administration and logistics, rather than command of individual Donkey Units.[32]

Donald A. Seibert, a former advisor of Donkey 3 and Donkey 13, later explained that American officers' control could only be exerted through the leaders of the guerillas, remarking, "In no sense did I command the pack." Each leader of a Donkey Unit recruited and led men from a particular region of North Korea with which he and they were familiar. Thus, many members of individual units had been friends before the war. This enhanced unit integrity and cohesion, but limited the influence of US advisors because the guerillas naturally associated themselves with their leaders. Most of the final decisions regarding unit operations/activities were made by the respective leaders, and advisors could merely suggest operations or discourage "questionable ones."[33]

REORGANIZATION OF PARTISAN OPERATIONS

In May 1951, the Eighth Army revised its organization for partisan operations which, since inception, had been under the direction of the Attrition Section of the G3 Miscellaneous Division. This had the undesirable result of a staff section engaging in operations, instead of Special Forces. Thus, on 15 May, the Attrition Section was deactivated, and reborn as the 8086th Army Unit, Miscellaneous Group.[34]

Other changes of name and command relationships followed. First, the Miscellaneous Group became the Guerilla Section under the FEC/Liaison Group (FEC/LG) (in Tokyo) and the FEC/Liaison Detachment, Korea (FEC/LD[K]) (in Taegu). On 10 December 1951, the Guerilla Section was renamed the 8240th Army Unit, FEC G-2, which, ultimately came under the operational control of the Combined Command for Reconnaissance Activities, Korea (CCRAK), 8242nd Army Unit, on 27 September 1952. Fortunately, throughout these many permutations, the focus remained on the guerillas.[35]

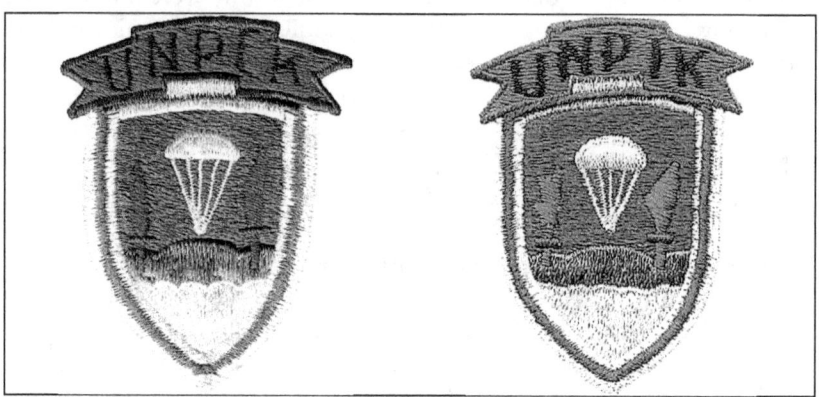

Unofficial shoulder patches worn by American advisors in the 8240th Army Unit. In November 1952, the guerillas of the 8240th AU were designated as United Nations Partisan Forces, Korea (UNPFK). This changed, in September 1953, when UNPFK was re-designated as the United Nations Partisan Infantry, Korea (UNPIK), to allow the awarding of the Combat Infantry Badge to the advisors.

During reorganization of U.S. Army oversight, partisans operating from the TF Leopard base on the west coast continued their missions. However, the beginning of the truce talks, in July 1951, had an impact on them. As the prospect of a major Eighth Army offensive faded, the static military situation meant that the partisans could not be optimally used in support of conventional military operations as a behind-the-lines irregular force. As pressure on the front line eased, North Korean and Chinese forces were freed for increased security measures within North Korea, posing a greater threat to the partisans. The morale of the irregular forces, who were fighting for a united non-Communist Korea, was also negatively affected by the realization that liberation of the north was increasingly unlikely. Nonetheless, west coast partisans continued to carry out offensive operations.[36]

12

Communist Spring Offensive

Photo 12-1

Shore fire control party from USS *Toledo* (CA-133) in an observation post overlooking the Han River, Korea, circa late April or May 1951. They are there to spot and correct the heavy cruiser's gunfire should enemy forces appear in the area.
National Archives photograph #80-G-432346

In April 1951, a major structural revision of Naval Forces Far East took place. Vice Admiral Andrewes who, for six weeks, had commanded Task Force 95, was relieved by Rear Adm. Alan K. Scott-Moncrieff, RN, as commander, Task Group 95.1, and command of the Blockading and Escort Force (TF 95) reverted to Rear Admiral Smith. With the departure of Andrewes from the theater, Task Force 95 was shifted from under commander, Naval Forces Far East (ComNavFE) to commander, Seventh Fleet.[1]

At the highest level of command, on 11 April, Gen. Matthew B. Ridgway succeeded Gen. Douglas MacArthur as commander in chief, United Nations Command. Gen. James Van Fleet, USA, arrived in Korea in mid-April, to replace Ridgway as commander, Eighth Army.[2]

Vice Adm. Harold M. Martin, USN, who had relieved Admiral Struble as commander, Seventh Fleet, on 28 March, assumed responsibility for the interdiction campaign. All heavy ships were absorbed into Task Force 77 (Aircraft Carrier Force) operating in the Sea of Japan off the east coast of Korea. Task Force 95 (composed, at various times, of the light carriers of Royal Navy and Royal Australian Navy; U.S. light carriers and escort carriers; two U.S. destroyer divisions; the ROK Navy; and surface units of other UN nations) continued its west coast Blockading and Escort Force duties. In the Yellow Sea, the carrier element continued to work over western Hwanghae Province, surface ships continued their missions of bombardment and patrol, and guerilla raiding forces were put ashore.[3]

The UN ground forces were now astride the 38th parallel, and the question of how far to press the advance again presented itself. Intelligence indicated that the Chinese intended to hold at the dividing line, while preparing for a major offensive in May. Since there was plenty of evidence, not least the Communist diligence in bridge repair (following strikes by aircraft and shore bombardment), to show that these preparations were being earnestly pressed, this intelligence was taken seriously. To hinder the enemy build-up and to maintain pressure on the Communist armies, Eighth Army began an advance across the parallel, Operation RUGGED, on 5 April.[4]

ROK PATROL FRIGATE COMES UNDER AIR ATTACK

In addition to the threat posed by the enemy's ground forces, its air strength, estimated in late March to have reached a total of some 750 aircraft, was growing stronger. Allied B-29 attacks on northern targets were meeting heavy MiG opposition; on 29 March, a twin-jet bomber was sighted over central North Korea; and enemy efforts were in progress to rehabilitate the North Korean airfields.[5]

For the naval forces, the danger was emphasized, on 15 April, when the ROK patrol frigate *Apnokkang*, lagging behind a force returning from the Yalu Gulf, was attacked by three enemy propeller-driven aircraft. Her crew shot down one of the attackers, but topside areas of the frigate were chewed up by strafing and near misses, and there were numerous casualties among her crew.[6]

Photo 12-2

Off the northwest coast of Korea, a guncrew, aboard the ROKN patrol frigate *Apnokkang*, man the 3-inch gun on her foc's'le, 26 April 1951. National Archives photograph 80-G-428919

ENEMY OFFENSIVE LAUNCHED

The immediate objective of the Communist Spring Offensive was the capture of Seoul. Their effort to do so began on the evening of 22 April, with a thrust down the center by the Chinese 20th Army. South of Kumwha, the ROK 6th Infantry Division collapsed under the weight of the attack and, as the enemy poured through the gap between the U.S. Army 24th Infantry Division to the west and the 1st Marine Division to the east, General Van Fleet ordered a withdrawal. Four days passed before the assault was checked. During this period, with the enemy out in the open, Fifth Air Force and carrier-based aircraft flew more than 1,000 close support sorties, inflicting very heavy casualties.[7]

On the 26th, the Communists launched their main effort in an attempted double envelopment of Seoul. One prong pushed down the Pukhan valley while, in the west, an attempt was made to ferry troops across the Han onto the Kumpo peninsula. Both efforts failed. The eastern threat to the capital was checked by the U.S. 24th and 25th Infantry Divisions, while on the Han, a day of strafing, by aircraft of the West Coast Carrier Element, reduced the enemy force crossing the river, to a level easily dealt with by the ROK Marine battalion defending the peninsula. Despite casualties estimated to be ten-fold greater than those suffered by UN forces, the Chinese gained no decisive advantage and, by 29 April, the front was stabilized once again.[8]

NAVAL AIR AND SURFACE SHIP SUPPORT

During the fierce ground combat, Task Force 77 carrier aircraft off Korea's east coast began, a ten-day sustained effort in support of the battleline, on 23 April. On the west coast, on the 26th, the threat to Seoul necessitated an evacuation alert at Inchon. The heavy cruiser USS *Toledo* (CA-133) arrived at Inchon to provide 8-inch gunfire support, as Rear Admiral Thackrey was called upon to take charge of exodus preparations. By 1 May, as evacuation shipping was beginning to arrive, some 200,000 refugees were gathered in the Inchon area.[9]

Photo 12-3

Heavy cruiser USS *Toledo* (CA-133) under way, location and date unknown.
Naval History and Heritage Command photograph #NH 67806

SECOND ATTEMPT BY CHINESE ARMY

The weight of the April enemy thrust toward Seoul led General Van Fleet to bolster his forces in the western lowlands. While this realignment of troops was in progress, the Chinese were shifting eastward to the central mountains where, on the night of 15 May, they attacked in strength. The brunt of the attack was borne by ROK divisions along the Soyang River, southeast of the reservoir. Again, the Communists broke through the line, and advanced 25 miles down the valley and across into the upper waters of the Hongchon area. Eastward, in the Sorak Mountains, enemy units overran the ROK III Corps and filtered down to the southeast; on the coast, the ROK I Corps withdrew south to Kangnung. In the west, Chinese divisions crossed the Pukhan below Chunchon and, on 17 May, opened a drive down the valley toward the Han.[10]

AMPHIBIOUS DEMONSTRATION

During the latter half of May, Rear Adm. Alan Scott-Moncrieff staged a fake amphibious landing at the bequest of Eighth Army, on a mainland beach opposite Cho-do Island. Believing that previous demonstrations had been too short and too transparent to evoke the desired reaction from the enemy, Scott-Moncrieff planned this with some finesse. Rumors of an impending landing were spread successfully by agents of Leopard Force. Aircraft from the British light carrier HMS *Glory*, flying cover for minesweepers working the approaches to the assault beach, reported a sign near the landing area which read "Welcome, U.N. Army." By 20 May, the preliminaries were completed and the heavy cruiser USS *Toledo* and Commonwealth ships were on hand to provide pre-landing fire support. That afternoon a dozen LCVPs, three loaded with Royal Marines and the others empty, landed the small number of troops opposite Cho-do. The Marines made a brief excursion inland, encountered no opposition, then re-boarded and the vessels departed.[11]

Photo 12-4

Drawing by Frank Norton of the coast between Cho-do and Sok-to islands. He was aboard HMAS *Warramunga*; a mechanism of hers is predominate in the foreground. Australian War Memorial photograph ART40023

These feints were popular with Army commanders. The enemy, after the events of the previous autumn, was fully aware of the amphibious capabilities of the U.S. Navy. Moreover, information from intelligence sources indicated that special pains were taken by the enemy to monitor the movements of the US Marines. But with the Marines still in the line at other locations, and given the slow reaction time of the Communist armies, there remained the question of whether the operation, of 20 May, would result in any diversion of Communist troops. Scott-Moncrieff was dubious, feeling that enemy communications were so poor that two or three days might pass before headquarters got the word.[12]

However, following the Cho-do affair, reports were received of troop movements across the Taedong River into previously undefended areas of Hwanghae Province. The passage of time also brought an increasing concentration of defensive forces, opposite Cho-do.[13]

ROLE OF TASK ELEMENT 95.11

On 20 May, the light carrier USS *Bataan* launched 12 combat air patrol and 37 offensive sorties—including 25 armed reconnaissance and 12 target combat air patrol aircraft in support of the amphibious feint in the Manap-to area, near Cho-do Island.[14]

Photo 12-5

Dutch destroyer *Van Galen*, on 22 October 1943, sporting her World War II pennant number G84, which was changed several times and was D803 during the Korean War. Australian War Memorial photograph 305848

Bataan had left Sasebo a day earlier, at 0700 on 19 May, with Rear Adm. Allan E. Smith (CTG 95) embarked and the Dutch destroyer HNMS *Van Galen* as escort, bound for the west coast of Korea. At 0945, the Australian destroyer HMAS *Warramunga* joined off the northern approach to Sasebo as an additional escort. The three ships arrived in the carrier operating area a little before noon. *Bataan* relieved HMS *Glory* in Task Element 95.11, and Capt. William Miller Jr., USN, the American light carrier's commanding officer, took over the duties of task element commander.[15]

Task Element 95.11 then comprised *Bataan*, the destroyer USS *Rupertus* (DD-851) with commander, Destroyer Division 32 embarked, the destroyer USS *Fechteler* (DD-870), and the destroyers *Van Galen* and *Warramunga*. Surface operations by the four destroyers involved supporting flight operations and replenishment, as well as radar surveillance of the northern part of the Yellow Sea at night. Formation 4R was used throughout, involving a circular screen during the day and an anti-submarine screen at night.[16]

Commencing on 22 May, radar patrol "Bugatti" was conducted nightly, to provide early warning of any enemy surface or air movement between the Shantung peninsula and Korea. The patrol had formerly been carried out by Task Element 95.12. In the new format conducted by TE 95.11, a destroyer was detached at the end of each day's flight operations to patrol north-south along longitude 124°00'E, between latitudes 37°40' and 39°00' North. After carrying out a prescribed search, the destroyer would rejoin the formation prior to resumption of flight operations the following morning. Only U.S. destroyers were used for these patrols, owing to their superior radar search capabilities.[17]

Air operations included armed-reconnaissance, close-air support, air-spot for shore bombardment, and combat air patrol missions. Armed-reconnaissance flights primarily involved the interdiction of enemy shipping—with emphasis placed on the destruction of junks and sampans in the Taedong Gang Estuary. Land transportation routes in the Hwanghae-do region were attacked when feasible, involving primarily, the systematic destruction of railway bridges, and the location and destruction of vehicle parks.[18]

Combat air patrol was maintained over Task Element 95.11 (carriers and escorts), and over UN forces conducting minesweeping and diversionary amphibious operations on the Chinnampo coast near Cho-do Island during the period 20-22 May. Armed-airspot was also provided for the diversionary forces and, when possible, was made an additional task of combat air patrol already assigned.[19]

At the request of commander, First Corps, Eighth Army, the Han River Estuary was kept under close air surveillance, to detect any activity that would indicate an enemy attempt to cross the estuary from the vicinity of Kaesong to the Kimpo peninsula above Inchon. No such activity was detected, and the surveillance was discontinued as unnecessary, after 26 May, because of the advance of UN ground troops north of the Han.[20]

Photo 12-6

Two North Korean soldiers near Kaesong on the 38th parallel, June 1950. Australian War Memorial photograph P00716.010

CHINESE OFFENSIVE HALTED

By 21 May, the Communists had been stopped all along the battle line. Despite a gain of 30 miles in the eastern mountains, and a considerable penetration in the Pukhan Valley toward the Han, nothing decisive had been accomplished, and enemy losses were higher than before. On the 23rd, Admiral Thackrey began to release shipping sent to Inchon. The evacuation alert ended two days later, and *Toledo* was at last relieved of her fire support duties.[21]

13

Royal New Zealand Navy Action

The unspectacular role of carrying personnel and supplies to Korea was perhaps the Navy's greatest contribution.

—Vice Adm. C. Turner Joy, commander, Naval Forces Far East.[1]

Photo 13-1

Drawing by Frank Norton of HMNZS *Rotoiti*, with the seas on her port bow, crossing the Straits between Korea and Japan, during the frigate's duty in the Korean War, 1952. Australian War Memorial photograph ART40055

New Zealand's participation in the Korean War involved six *Loch*-class frigates, HMNZS *Pukaki*, *Tutira*, *Rotoiti*, *Taupo*, *Hawea* and *Kaniere*. A measure of the significant contribution of this sparsely-populated nation was that the ships' crews comprised about half the manpower of the Royal New Zealand Navy.[2]

When the first of her ships—HMNZS *Tutira* and *Pukaki*—arrived in theater on July 1950, they were assigned escort duty, with up to four convoys a week shepherding transports and supply vessels. The frigates' task, shared by ships of other navies, was generally considered the most thankless of the sea war—dull, daily routine patrol. Escort of individual convoys was pared down as the war progressed, and other vital requirements presented themselves. Newly arrived ships in theater were often tasked with relieving those engaged in escort work, as the first duties in order to familiarize themselves with the area.[3]

SUPPORT OF ROK NAVY MINESWEEPING FORCES

> *Frustration and difficulty have never been absent from the work and its progress is due to the determination, hard work and ingenuity of the officers and crews of the minesweepers. Woefully short of all their needs and lacking even the elementary tools for splicing, they have improvised and shared, never admitting they were beaten until they reached a stage where they could only muster enough equipment to stream one double Oropesa sweep.*
>
> —Lt. Comdr. Brian E. Turner, commanding officer of HMNZS *Rotoiti*, praising the efforts of the South Korean minesweepers *Guwolsan* (YMS-512) and *Ko Yung* (YMS-515) off Haeju. Korea.[4]

Photo 13-2

Yard minesweeper USS *YMS-413*, before her transfer to the ROK Navy for service as *Kwang Chu* (YMS-503), location and date unknown.
Courtesy of NavSource and David Richter

Frigates (like minesweepers) are "maids of all duties." Equipped with smaller naval guns than destroyers and cruisers, they were not ideal candidates for heavy shore bombardment and were, thus, employed for myriad other duties. In spring 1951, HMNZS *Tutira* and *Rotoili* served as "shotguns," protecting ROKN minesweepers from enemy attack, while these vessels carried out their dangerous work.[5]

Between 15 March and 7 April 1951, first HMNZS *Rotoiti*, then *Tutira*, supervised operations by the South Korean Navy minesweepers *Kyong Chu* (YMS-502) and *Kwang Chu* (YMS-503). The South Koreans worked hard in spite of many challenges facing them. Their ships were long overdue for much-needed maintenance, and aging equipment caused frustrating delays. They were always short of provisions and were only sustained by fuel and food from the *Tutira*. Despite these difficulties, there was tangible evidence of success; two moored mines were cut, rose to the surface, and were destroyed by rifle fire.[6]

HMNZS *Rotoiti* had a double role in the minesweeping. When relieved by *Tutira*, she supported two other ROKN minesweepers, *Guwolsan* (YMS-512) and *Ko Yung* (YMS-515) off Haeju. This was the RNZN's first close working contact with the South Korean Navy, and relations between the sailors of both were amicable.[7]

Following her support of minesweeping operations, *Tutira* arrived at Kure, on 26 April 1951, and handed off to sister ship HMNZS *Hawea*, which also arrived there that day. *Tutira* then underwent a short refit. At completion, she departed for Auckland, reaching home on 30 May.[8]

ACTION ASHORE IN APPROACHES TO CHINNAMPO

Photo 13-3

New Zealand frigate HMNZS *Rotoiti* (F625), location and date unknown. Courtesy of Torpedo Bay Navy Museum (photograph AAR0032)

After minesweeping support, patrol duties in the Yellow Sea followed for the New Zealand frigates remaining in Korean waters. First was *Pukaki* with *Rotoiti* relieving her off the west coast of Korea to take up her first patrol. Their duty was to safeguard the "Cigarette Route," and to deny North Korean forces access to the off-shore islands and South Korea's coastline and ports. These patrols, often in adverse weather conditions, involving much observing and waiting, were long and boring. They did, however, offer frequent contacts with other UN Fleet elements, including battleships, aircraft carriers, cruisers, destroyers, frigates, minesweepers, and South Korean patrol craft.[9]

Pukaki returned to New Zealand in December, completing her only deployment, and was paid off into reserve, on 22 December 1950.[10]

In an effort to strike at the enemy, by gaining intelligence ashore, HMNZS *Rotoiti*'s commanding officer, Lt. Comdr. Brian Edmund Turner, formed a landing party comprised of his fittest young sailors. When the frigate was next in port at Kure for crew rest and ship maintenance, the chosen members underwent a short period of training with the 41 Independent Commando, Royal Marines. The week-long instruction was devoted mainly to stealth and concealment, armed and unarmed combat, and taking of prisoners.[11]

The 41 Independent Commando had been formed at Bickleigh, in August 1950, specifically for the war in Korea. Specializing in carrying out amphibious raids on the Korean coastline, the Commando was attached to the 1st US Marine Division when the Chinese entered the war, and took part with this division in the breakout from the Chosin Reservoir in December 1950. This commando unit, disbanded in February 1952, was later awarded a United States Presidential Unit Citation, in 1957, for services with the 1st Division, USMC.[12]

41 Independent Commando Royal Marines

Commanders:
Lt. Col. Douglas B. Drysdale DSO, MBE RM
Lt. Col. Ferris Nelson Grant, CB RM

Formed:
16 August 1950 at Bickleigh Camp
(Commando School at South Hams, UK)

Disbanded:
22 February 1952 at Stonehouse Barracks
(Plymouth, UK).

FIRST RAID

On 11 July 1951, not long after *Rotoiti* resumed her "Cigarette Route" patrol activities, Lieutenant Commander Turner decided to carry out a quick daylight attack against an isolated enemy observation post. The purpose of the midday raid was to capture prisoners for intelligence purposes. Under cover of shore bombardment by *Rotoiti*'s 4-inch gun, and Bofors 40mm AA guns, her motor boat beached at Sogon-ni Point, in the approaches to the port of Chinnampo.[13]

Scrambling ashore from the boat was a two-man assault team, Able Seamen Norman Scoles and Jim Button, armed with sub-machine guns and grenades, and a command and cover party armed with Bren guns and rifles. After the bombardment ceased, the assault pair scaled a small cliff to attack the North Koreans manning the post. A brief action left one enemy soldier dead and two taken prisoner. The whole party, including captives, then withdrew, under cover of additional bombardment, and rejoined *Rotoiti*. The action took less than an hour.[14]

Photo 13-4

Weary soldiers of B Company, 1st Battalion, The Royal Australian Regiment (1RAR), just back from an all-night patrol during which they engaged an enemy force. Left to right: Lt. Jack Harold Skipper, Lance Corporal David Young, and Corporal Bill Crotty. All three are armed with Owen sub-machine guns.
Australian War Memorial photograph HOBJ3526

RECOGNITION OF VALOUR
Four individuals involved with the action ashore against the enemy, initiated by Lieutenant Commander Turner, received awards for valour:

Distinguished Service Cross	Distinguished Service Medal
Lt. Comdr. Brian E. Turner, RNZN	Able Seaman Norman J. Scoles, RNZN
Lt. Richard S. F. Webber, RN	Able Seaman Edward J. Button, RNZN

SECOND RAID
Five weeks later, on 26 August 1951, another raid involving the New Zealanders took place in the same Sogon-ni locality. Unlike Turner's independent action, this operation involved higher authority and greater resources, as evidenced by the participation of Royal Marines from HMS *Ceylon*, and gunfire support from the cruiser. As before, the mission was to capture enemy personnel for intelligence purposes.[15]

The assault party under Lt. John D. Hunter, a Royal Marine officer from *Ceylon*, included three Marines—Corporal McGregor, Noel Harker, and Cecil Self—and three *Rotoiti* "Commando" seamen. Scoles, a veteran of the first raid, was an obvious choice. He was joined by Able Seaman Robert Marchioni (a replacement for Button, who had departed the ship for a new duty assignment), and Stoker Mechanic 'Buster' Dunn. Dunn, like Marchioni, had not participated in the first raid.[16]

Executed in pitch darkness, both the landing and the mission proved to be more difficult than the daylight snatch of the previous month. At fifteen minutes past midnight, the two boats carrying the raiders left *Rotoiti*'s side, about six miles from shore. Things did not go smoothly, owing to low water. The tide was out, resulting in the boats grounding some distance from the beach, and the assault party and 13-man cover group were required to wade ashore.[17]

Once ashore, the cover party, under the command of Lt. Richard Webber, Royal Navy, quickly ensured that the beachhead was safe, while the assault party climbed up a gully to a shelf behind the beach. From there, they moved north along the slightly sloping top, to approach the enemy position a short distance away. Concealed by long grass, the raiders lay down and studied what Hunter later described as "a sort of pillbox." Three enemy about thirty feet away were looking out to sea. Just prior to and during the initial stages of the raid, *Ceylon*, positioned about ten miles north of the assault site, opened fire with her main armament to create a spectacular display of star-shell bursts, intended to distract the enemy's attention.[18]

Hunter sent McGregor, Harker, and Dunn to the right to cover the others as they approached the enemy position. Hunter, Self, Scoles, and

Marchioni then got up and moved forward. Surprise was lost when a noise alerted the North Koreans. After an unanswered verbal challenge by the enemy, a long continuous burst of machine-gun fire directed at the assault group followed. The members of the assault party dove to the ground, but the first few rounds later proved lethal to Marchioni, who was severely wounded. Reacting quickly, Scoles half-raised up and threw a grenade. Its detonation, and fire from the cover party on the beach below, suppressed the enemy fire.[19]

Hunter's orders directed him to withdraw if confronted by serious opposition. Faced with a strong enemy group of unknown size in the darkness, and with a man wounded, he did not hesitate. He and Scoles dragged Marchioni to the cliff and went over it. By the time the whole assault group met up on the beach below, Marchioni had died. Harker carried his body along the beach for a while, Corporal McGregor then took over, and later Scoles took his deceased countryman.[20]

Things became desperate when the group came to a rocky headland that separated them from the section of beach where the boats were grounded. With North Koreans running along the cliff above, Hunter feared that unless his party moved quickly, they would be cut off from the boats. Carrying Marchioni's body around or over the rocks would increase the danger, and Hunter decided to leave Marchioni's body. After placing it in a niche in the rocks, the group struggled past the headland to the boats a short distance away. The assault party quickly boarded, followed by the cover party.

Photo 13-5

Bren gun practice by an Australian serviceman during World War II, 25 June 1940. Australian War Memorial photograph 002017

The Bren gunners and the Marines, then in the boats, kept the enemy soldiers on the top of the cliff under fire, as the boats backed clear. Once they were safely away, both warships shelled the back of the beach. The raiders arrived at *Rotoiti* at 0215. One Marine recalled that aboard her, there were "glasses of rum all round but also a tot of sadness," as they celebrated the conclusion of the dangerous mission, and grieved the loss of Marchioni.[21]

Lt. Comdr. Brian E. Turner, HMNZS *Rotoiti*'s commanding officer, sought permission the next day to recover Marchioni's body and bring him home. This request was withheld by the higher authority and a possible disaster averted. Instead, Padre Harry Taylor conducted a Memorial Service on *Rotoiti*'s quarterdeck, within sight of the place where Able Seaman Robert Marchioni died in the service of New Zealand.[22]

Marchioni was the only Royal New Zealand Navy combat casualty of the Korean War.[23]

14

Ground War at an Impasse, Armistice Talks Begin

Photo 14-1

United Nations' delegates Vice Adm. C. Turner Joy, USN (left), Maj. Gen. Henry I. Hodes, USA (center), and Rear Adm. Arleigh A. Burke, USN (right), on the steps of UN House at Kaesong, Korea, 19 July 1951.
National Archives photograph #80-G-431923

On 23 June 1951, the Soviet delegate to the United Nations, Jacob A. Malik, proposed ceasefire discussions between the protagonists in the Korean War. Armistice discussions commenced at Kaesong, on 8 July. The chief of the UN Delegation was Vice Adm. C. Turner Joy, USN, commander, Naval Forces Far East. The other U.S. Navy delegate was Rear Adm. Arleigh Burke, commander, Cruiser Division Five.[1]

The combination of President Harry Truman's dismissal of General MacArthur as commander in chief, United Nations Command, and commencement of armistice talks resulted in the ground war becoming immobile. By mid-June, the battle line had been stabilized on favorable ground, and United Nations activity was largely limited to patrolling and to the improvement of defensive positions. For the next two years, as hopes of peace continued to be frustrated by lack of progress in the talks, the UN Air Force and the Navy principally shouldered the burden of offensive operations.[2]

BRIEF DUTY OF COLOMBIAN FRIGATE

The composition of the UN Naval Forces operating off both coasts of Korea, changed from time to time, as individual ships reported for duty, or left the theater, after their war tours were completed. On 5 May 1951, ARC *Almirante Padilla* (the first of three Colombian frigates to serve in the war) arrived at Sasebo, Japan, and was assigned to Task Element 95.13 (Patrol and Escort Element) operating off the west coast of Korea.[3]

Before discussing the *Almirante Padilla* Korean duties, it is worthwhile to note the tremendous effort required of her ship's company in bringing her to the Far East. On 6 September 1950, Colombian president Laureano Gómez Castro pledged a frigate to the UN Naval Command. At that time, the entire Colombian Navy consisted of two Portuguese destroyers (captured during the Colombia-Peru War of 1932), the ARC *Almirante Padilla* (ex-USN *Tacoma*-class patrol frigate purchased in 1947), and ten river gunboats.[4]

On 1 November 1950, the *Almirante Padilla*, with a crew of 190 (ten officers and 180 men), departed Cartagena (on the northern coast of Colombia) bound for San Diego, for combat refitting. Her crew knew that neither they, nor their ship were then ready for combat, but these were not all of the frigate's limitations. Lt. Comdr. Julio Cesar Reyes Canal, her commanding officer, later recounted, "Much to my surprise, two hours after leaving Balboa, Panama Canal Zone, for San Diego, I asked for fifteen knots. I was speechless when my chief engineer told me that the machinery was too bad and that we could only make ten knots."[5]

By the time *Almirante Padilla* arrived in San Diego, on 13 November, it was apparent that the state of her ship propulsion, communications, armament, and fire control were inadequate. Work at the Long Beach Navy Yard began, on 12 December. Although completion of an overhaul made the frigate seaworthy, new guns and fire control systems were needed for impending combat operations. A refit of her weapons

Ground War at an Impasse, Armistice Talks Begin   141

systems was so expensive that President Gómez had to personally authorize the work.[6]

Almirante Padilla finally made it to Sasebo in spring 1951, and was assigned to Task Element 95.13. She carried out coastal blockade patrols with the cruisers HMS *Ceylon* and *Kenya*, destroyer HMCS *Sioux*, and patrol frigate USS *Glendale*. Because shallow waters along the coast denied movements of larger warships, the Colombian and American frigates, and South Korean minesweepers, conducted the inshore patrols. On 14 June 1951, *Almirante Padilla* was ordered to the east coast to join a siege of Wonsan initiated by Rear Admiral Smith, in February.[7]

Almirante Padilla's subsequent duty was on the east coast (as was that of the other Colombian frigates which served after her). During her final two patrol periods, *Almirante Padilla* shelled Wonsan targets, sunk numerous mines with gunfire, rescued several downed UN pilots, and supported a SMG (Special Mission Group) intelligence collection force sent to the island of Yang-do, near Songjin. Her combat service ended, on 12 February 1952, upon her relief by ARC *Capitán Tono*.[8]

SALVAGE OF SOVIET-BUILT MIG AIRCRAFT

Photo 14-2

Russian built Mig-15 jet aircraft piloted by North Korean pilot Lt. No Kum Sok, shortly after he landed the aircraft at the No 77 Squadron RAAF base at Kimpo, South Korea, on 21 September 1953, and surrendered the aircraft to United Nations forces.
Australian War Memorial photograph P05112.001

On 9 July 1951, word was received from the UN Joint Operations Command, that a Soviet-built MiG was down in shoal water off the mouth of the Chongchon River. The escort carrier USS *Sicily*, which arrived as relief for USS *Bataan*, was ordered to search for the fighter aircraft, and American officers in charge of "Leopard" guerillas on Paengyong-do and those of "Salamander" on Cho-do Island, were alerted.[9]

Two days later, planes from HMS *Glory* located the MiG (in shallow water a couple of miles offshore, and thirty-three miles north of the estuary of the Taedong River) fifteen miles from the reported position. Finding the MiG had been challenging; the weather was foggy and the aircraft awash (visible on the surface of the sea) only at low water.[10]

A recovery effort was quickly initiated, owing to the importance of technical intelligence associated with the enemy aircraft. Salvage of the plane was considered risky. Its location was behind enemy lines (less than 10 minutes flying time from the Chinese Antung airfields at which scores of MiG-15s were based), and navigationally difficult. However, ComNavFE ordered that "every effort" be taken, and Captain Lloyd-Davies, commanding officer of HMS *Ceylon*, worked out a plan.[11]

On 18 July, the dock landing ship USS *Whetstone* (LSD-27) arrived at Cho-do from Inchon, carrying, in her well deck, a utility landing craft (LCU) equipped with a special crane. Operations the next day, ended with the LCU fast on a sandbar, but later extracted. Efforts continued on the 20th, with HMS *Glory* providing air cover and the frigate HMS *Cardigan Bay* on hand for fire support.[12]

Photo 14-3

Dock landing ship USS *Whetstone* (LSD-27) off Point Loma, California, in the 1950s. U.S. Navy photograph courtesy of NavSource and David Buell

A U.S. Navy helicopter operating from the British carrier buoyed the aircraft site to mark its location, while *Glory* aircraft led the LCU through sandbars offshore, then up a narrow approach channel to the site, an area of mudbanks and treacherous tides. *Cardigan Bay*'s motorboat accompanied the landing craft. By evening the plane's engine had been recovered, as well as major portions of its fuselage, using the LCU crane at low water. Smaller fragments were retrieved by shallow water divers.[13]

The following day, 21 July, the landing craft returned to the site to recover remaining small pieces of the aircraft, accompanied by the frigate's motor boat and also, this time, her whaler and dinghy. In the afternoon, *Sicily* pilots sighted thirty-two MiGs heading for the area, but foggy weather shielded the operation from view, and no trouble ensued. On the 22nd, the LCU and its cargo were embarked in USS *Epping Forest* (LSD-4), a sister ship of the *Whetstone*. The MiG wreckage was brought back to Inchon, and later transferred to Wright-Patterson Air Force Base (Ohio) in the United States, for analysis.[14]

Photo 14-4

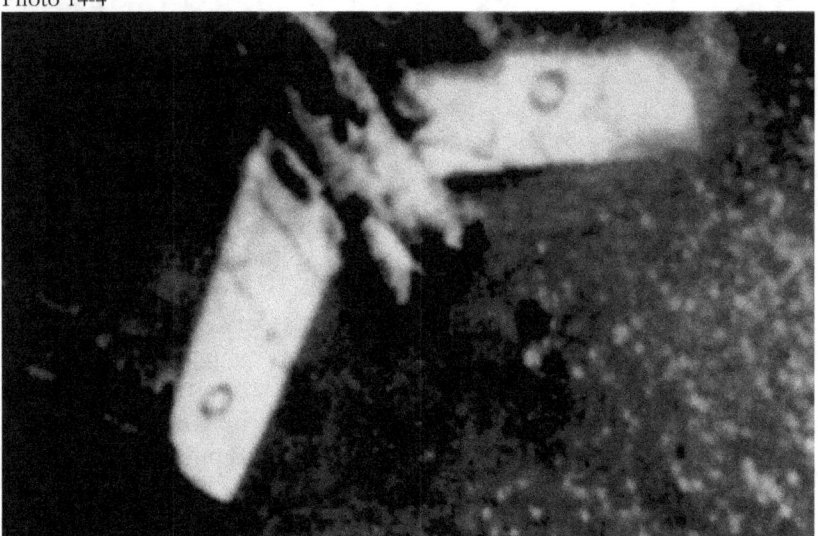

Wreckage of the Mikoyan-Gurevich MiG-15 fighter before recovery from the shallows. *Naval Aviation News* December 1951

DEBATE ABOUT LINE OF DEMARCATION

In late July 1951, the delegates at Kaesong took up the question of a demarcation line between North and South Korea. The Communists, who had suffered a net loss of territory, insisted on the 38th parallel.

UN negotiators, for their part, sought compensation in a line north of the existing front. From this discussion, arose the question of who controlled the territory of the Yonan and Ongjin peninsulas. A part of the Hwanghae Peninsula, they lay south of the 38th parallel and west of the Imjin River, which ran southwestward from Haeju.[15]

Map 14-1

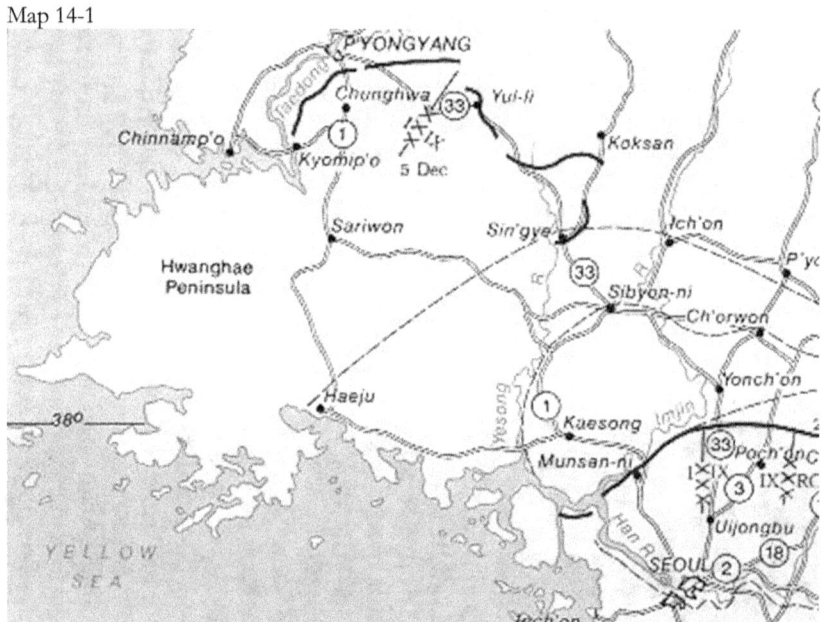

Yonan and Ongjin peninsulas lay south of the 38th parallel in Hwanghae Province
Source: https://campsabrekorea.com/maps-of-korea-1950s--present.html

Largely untouched by war, and but lightly defended by the enemy, the coastal areas of Hwanghae Province were vulnerable both to Communist movements from the sea, as well as to raids by guerillas operating from the offshore islands. On the evening of 24 July, as the question of the demarcation line arose, Vice Admiral Joy contacted commander, Task Force 95, Rear Adm. George C. Dyer, USN, to ask for a show of strength in the Han River Estuary as close as possible to Kaesong, where the talks were being held. Dyer (who had succeeded Rear Admiral Smith on 20 June in command of Task Force 95) committed all but one of his west coast frigates to the operation at once. HMS *Glory* was ordered from Sasebo to join *Sicily*, and a check mine sweep of the entrance to Haeju Bay was undertaken before entry by the warships.[16]

HAN RIVER NAVAL DEMONSTRATION

The Han River demonstration was a very difficult naval operation.... My hat is off to the British Navy and the Commonwealth Navies for the courage, tenacity, and high degree of seamanship they showed in accomplishing this job. When they reached the Kyodong Island area, we established an anchorage there, and commenced taking the enemy under fire. The survey then proceeded both westward and eastward. However, the only navigable channel found was one that went westward along the north of Kyodong Island, then turned north at Inson Point and proceeded to the eastward.

As soon as we showed up north of Kyodong Island, the enemy started constructing batteries at Ayang Point and at the mouth of the Yesong River. There was a railroad line that ran from Yenan to Kaesong, and a ferry across the Yesong River. There was heavy traffic on this ferry. To shell it regularly, the frigates had to get up to the mouth of the Yesong. The enemy would plant machine guns and mortars in the rice fields at night, and when the frigate came along in the morning, would shell the frigate, and there would be a close fight.

—Rear Adm. George C. Dyer, USN (commander, Task Force 95), describing the Han River Naval demonstration by UN forces, undertaken to counteract the Communists' immediate claim made at the truce talks, that the 200 square miles south of the 38th parallel and west of the Imjin River were in their hands.[17]

On 25 July, three frigates—HMAS *Murchison*, HMS *Cardigan Bay*, and ROKN *Apnokkang*—initiated operations to probe the Han River to the limit of navigation as a demonstration of Allied control of the area. The reason for this action was to contest the Communists' assertion that they controlled the area. Since the city of Seoul was located at the headwaters of the Han River, it was imperative that any cease fire agreement not surrender the maritime approaches to the South Korean capital to Communist control.[18]

The preliminary attempt by these ships to negotiate the uncharted waters of the Han River approaches was aborted. Currents in the Han ran from 4-10 knots, the navigable channel shifted rapidly from one side of the river to the other, there were no navigational aids (such as a buoy system), and the tides ran from 12-25 feet. The frigates returned to the assembly anchorage, and were joined by the frigate HMS *Morecambe Bay* and six craft of the South Korean Navy.[19]

On the evening of 26 July, the entire group (Task Unit 95.12.2) proceeded and anchored in the chosen position (37°48.7'N, 126°20.6'E) from which gunfire could be directed over a wide arc on the north shore of the Han River. This point was in an area 30 miles in from the Yellow Sea, where the river widened into a bowl about four miles in diameter, between the Communist-held north bank and the South Korean-controlled south bank. The area where the ships anchored came to be known as the "Fork," because channels radiated from it between mud banks, which the frigates used for reconnaissance and close attack. From this advanced anchorage, eight miles west of Kaesong, the ships' guns dominated a large area of the north bank of the Han. Eastward, attack was limited by the neutral zone around Kaesong. But to the west, troop concentrations, dumps, gun positions, and railway tracks as far inland as the city of Yonan, were within range of naval gunfire.[20]

Photo 14-5

Painting by Ken McFadyen of the frigate HMAS *Murchison* shelling North Korean positions on the Han River Estuary, near the mouth of the Yesong River, Korea, 1951. Australian War Memorial photograph ART40820

From 27-29 July, the heavy cruiser USS *Los Angeles* (CA-135), after making her way up the swept channel into Haeju Bay, shot up targets on its western shore. Meanwhile, the Commonwealth frigates shelled the northern bank of the Han River Estuary. HMAS *Murchison* remained with this group, engaged in bombarding shore installations, troop concentrations, gun emplacements, and store dumps, until 4 August. Naval gunfire was maintained day and night with planes from Task

Element 95.11, assisting during the day by spotting and reporting the fall of the rounds.[21]

By early August, bombarding ships had succeeded in penetrating upstream to fire on the city of Yonan from the southeast and northward up the Yesong River. On the 17th, three of the frigates found 400 enemy troops along the river bank and took them under fire. Late in the month, a survey of the Han was begun by a UDT detachment aboard the high-speed transport USS *Weiss*, and the channel was buoyed by the fleet tug USS *Abnaki*.[22]

Over the next several weeks, ships came and went. A total of twelve Commonwealth destroyers and frigates, the U.S. Navy transport ship and fleet tug, and ROK Navy frigates and patrol craft participated in the long, dangerous Han River operation. These vessels, all under the command of Scott-Moncrieff (CTG 95.1), are identified in the table.

Rear Adm. Alan K. Scott-Moncrieff, RN
Task Unit 95.12.2 (Han River Demonstration)

Royal Australian Navy	Royal Navy
Frigate HMAS *Murchison* (F442)	Destroyer HMS *Comas* (D43)
Royal New Zealand Navy	Frigate HMS *Amethyst* (F116)
Frigate HMNZS *Hawea* (F422)	Frigate HMS *Black Swan* (F57)
Frigate HMNZS *Rotoiti* (F625)	Frigate HMS *Cardigan Bay* (F630)
Frigate HMNZS *Taupo* (F421)	Frigate HMS *Morecambe Bay* (F624)
United States Navy	Frigate HMS *Mounts Bay* (F627)
Fleet tug USS *Abnaki* (ATF-96)	Frigate HMS *St. Brides Bay* (F600)
High-speed transport USS *Weiss* (ADP-135)	
Republic of Korea Navy	
Frigates and patrol craft[23]	

HMAS *Murchison* returned to the Han River operation several times that summer and autumn. While aboard, on 28 September 1951, for a tour of the estuary, Rear Admiral Dyer witnessed, first-hand, a particularly hot action, as described by Lt. Comdr. Allan N. Dollard, RAN, the ship's commanding officer:

> About 1600, unsuspected batteries of 75mm guns, 50mm guns, and mortars opened fire on us from the north bank of the Han. We had just reached the Yesong River and had dropped our anchor to let the current turn us around, when the first mortar [round] hit.[24]

Dollard quickly weighed anchor, and maneuvered *Murchison* clear of the vulnerable position, while her 4-inch guns engaged in counterbattery fire, scoring several hits and silencing the enemy.[25]

Photo 14-6

Gun crew of one of HMAS *Murchison*'s 4-inch guns, during combat operations in the Han River Estuary, circa 1951. Australian War Memorial photograph P03069.001

ARMISTICE TALKS SUSPENDED, RESUME AFTER CONFERENCE SITE RELOCATED TO PANMUNJOM

By this time, the optimism which had accompanied the opening of armistice talks was gone. In early August, General Ridgway briefly suspended negotiations; and late in the month the Communists, in turn, refused to talk. Only in late October, with transfer of the conference site to Panmunjom, were fully-attended sessions resumed.[26]

The Han River demonstration lasted until 27 November 1951, at which time the negotiators agreed upon a provisional ceasefire line, with the disputed territory recognized as in UN hands.[27]

AWARDS FOR VALOUR

HMAS *Murchison*'s commanding officer and one other of her officers, received the Distinguished Service Cross. The Australian frigate spent more time inside the Han Estuary than any other ship, often her naval percussion on the river was audible at the ceasefire conference tables. In tribute to his coolness during many "warm moments," a British admiral dubbed Lt. Comdr. Allan Dollard, "Baron of the Han." Commodore Dollard retired in early 1973, after 41 years' service (beginning in 1932) to the Royal Australian Navy.[28]

Distinguished Service Cross
Lt. Comdr. Allan Nelson Dollard, RAN
Lt. James Maxwell Kelly, RAN

15

Enemy Threatens Northwest Islands

Photo 15-1

Drawing by Frank Norton of the British cruiser HMS *Ceylon* as seen from the Australian destroyer HMAS *Warramunga*, off Sok-to Island in the Yellow Sea, Korea, 30 June 1952. Australian War Memorial photograph ART40019

With an agreement for a provisional ceasefire line emerging from the armistice talks, frigate bombardment in the Han River was terminated, and action on the front lines tapered off into patrolling and fortification of defensive positions. But as the ground battle became largely static, the enemy began an offensive effort in offshore islands, concentrating on those off northwest Korea.[1]

On the east coast, except for those in Wonsan harbor, there were only four islands of importance. Of these, the two largest—Mayang-do on the 40th parallel and Hwa-do off Hungnam—were enemy-controlled. Northeast of Songjin, the Yang-do island group, two miles offshore, hosted Allied intelligence personnel moving in and out of

North Korea, and would later become an ROK Navy PT (motor torpedo boat) operating base. Nan-do, a little island on the 39th parallel, hosted Task Force Kirkland, an Eighth Army unconventional warfare organization which controlled partisan activities on the east coast.[2]

On the west coast, along southern Hwanghae province, from the Han River Estuary to Korea's western tip, numerous coastal islets were employed as bases by partisan groups, of which Leopard Force was the most prominent. Off the Chinnampo approaches, Sok-to and Cho-do supported guerilla and clandestine operations; the U.S. Air Force desire to install radar facilities and rescue helicopters on the latter island, waited only on improved security. To the northward in the Yalu Gulf, a group of islands (seized by the ROK Navy in November 1950), were also home to numerous groups of anti-Communist guerillas.[3]

U.S. ARMY AND FIFTH AIR FORCE "SPEC OPS"

The number of independent military organizations on these islands, and their use of some of the same coastal waters, led at times to situations of considerable complexity. U.S. Army, Navy, and Air Force, along with UN naval units, in addition to anti-Communist partisans were involved.[4]

"Line-crossers" were Koreans recruited to gather clandestine human intelligence and were inserted behind enemy lines by parachute or other means. Agents who took on this hazardous duty, once in, had to exfiltrate through enemy lines to bring back their intelligence. Following unacceptable losses of these agents, an expanded program of espionage, code named "Salamander," was implemented by U.S. Army 442nd Military Intelligence Detachment (CIC). This expanded program utilized teams from the 3rd and 25th Infantry Division to transport agents by small boats around the enemy's flank on the west coast. Included in this effort was the use of Korean-manned fishing boats, launched from west coast islands, which were in the hands of anti-Communist North Korean partisans, and rendered, by the UN naval blockade, more or less secure from hostile attack.[5]

The "Leopard" command, as noted previously, involved oversight of guerilla operations. Leopard at Paengyong-do cooperated well with the blockading force, as did Salamander at Cho-do.[6]

Salamander operations mounted from Paengyong-do, just below the 38th parallel, soon after the launch of the expanded program moved forward to a base at Cho-do, strategically located just a few miles off the North Korean coast. Its position gave the agents access to the west coast of Korea up to the Yalu River.[7]

22nd Crash Rescue Boat Squadron   **U.S. Fifth Air Force**   **6004th Air Intelligence Service Squadron**

Other groups were less accommodating about keeping naval forces apprised of their activities. In the same area off Korea's northwest coast, the U.S. Air Force 22nd Crash Rescue Boat Squadron, operated its own private navy—85-foot crash boats—for the rescue of downed pilots around the Cho-do area. Stationed far north to aid combat pilots ditching in the sea or even downed airmen making the shoreline from farther inland, the wooden-hulled craft were seen by some as useful transport for other purposes, including intelligence gathering.[8]

Crash boat crews drafted to support operations of U.S. Fifth Air Force 6004th Air Intelligence Service Squadron, Detachment 2, behind enemy lines, soon learned that the unit was a lot more than a "covert collection agency." Its members, who seemed to need a lot of guns to collect whatever it was they wanted, were charged with examining enemy materiel, and interrogating enemy deserters.[9]

Detachment 2 personnel were involved in the salvage of the MiG aircraft, discussed in the previous chapter. Included in a chronology of significant events in USAF's operations, compiled by the Air Force Historical Research Agency to commemorate the Korean War, is the following entry:

> July 21: A detachment of the 6004th Air Intelligence Service Squadron completed a week-long effort near Cho-do Island to recover the most components ever salvaged from a MiG-15 aircraft. This combined operation involved 5th Air Force aircraft providing high cover, British carrier aircraft flying low cover, and the U.S. Army contributing a vessel outfitted with a crane.[10]

It is understandable that Navy ships involved in this sensitive operation would not cite in their reports, the involvement of this clandestine unit. Based on the quoted material's reference to a "week long effort," the 6004th personnel arrived on the scene three days in advance of USS *Whetstone*. Perhaps the foggy conditions, that delayed discovery of the MiG, persisted, and Air Force members misidentifying *Whetstone* as an Army vessel may be forgiven their transgression.

MERGING OF RECONNAISSANCE ACTIVITIES

On 28 November 1951, representatives of all USAF intelligence gathering organizations in Korea met at Far East Command, Liaison Division, to discuss how to coordinate their activities. Air Force captain Donald Nichols represented Detachment 2, 6004th Air Intelligence Service Squadron. The conference resulted in the establishment of the Combined Command Reconnaissance Activities, Korea (CCRAK), a United States Air Force special operations unit.[11]

INCREASED THREAT TO FRIENDLY ISLANDS

Rear Admiral Scott-Moncrieff, after learning of the concentrated vessel traffic in his coastal waters, ordered that all unidentifiable travelers be apprehended and detained. By autumn this situation had improved, making it easier to quickly monitor whether an unidentified vessel plying coastal waters was friend or foe. As possession of islands off northwest Korea was linked to discussion at the armistice talks, regarding location of the cease-fire line, the enemy was showing interest in UN/guerilla-held islands. This interest added to the vulnerability of losing these islands, as the talks were adversely affecting the morale of the guerillas. They were fighting for a free North Korea, which now seemed less and less likely.[12]

CHO-DO – SOK-TO AREA, WEST OF CHINNAMPO

Two islands of particular value to UN forces were Cho-do and Sok-to. Located in the approaches to the Taedong Estuary, they guarded the northern end of the only deep-water channel along the northwest coast of the Hwanghae Promontory—the so-called "Cigarette route." Cho-do (located at 38°32'N, 124°50E) was a rocky, mountainous island, some six miles across at its widest point. It was situated only 3 ½ miles from the mainland, separated by a six- to nine-fathom deep channel. The much smaller Sok-to, lay about seven miles to the northeast of Cho-do, and less than two miles from the peninsula of Pip'a-got, which jutted out into the sea between the two islands.[13]

Due east of Sok-to was another peninsula, Amgak, and between it and Pip'a-got, a shallow bay containing the tiny islands of Chongyang-do and Ung-do. These islands (not shown on the map) were held by anti-Communist guerillas, and were particularly vulnerable to attack because, at low water, it was possible for enemy soldiers to cross on foot from the mainland over the exposed mud flats.[14]

Map 15-1

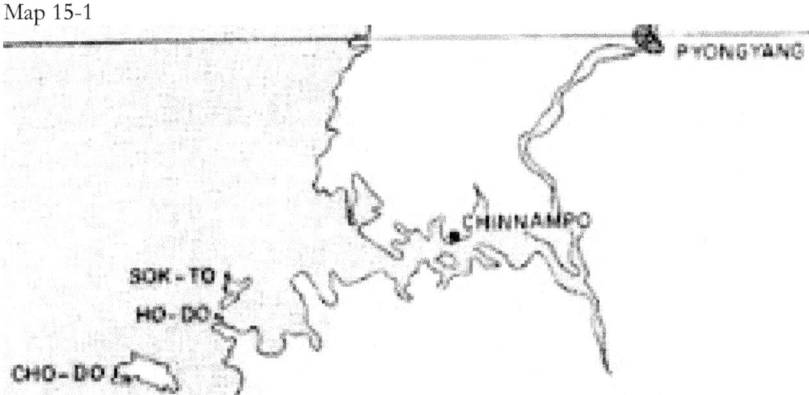

Key islands located west-southwest of Chinnampo, North Korea

Stationed on Cho-do and Sok-to were small detachments of well-disciplined, but poorly equipped ROK Marines usually under American command. Their chief duty was to maintain law and order on the islands and protect them against invasion from the mainland. Also based on the islands, but not under ROKM command, were large numbers of poorly-armed guerillas, nominally under American control. They were available to help defend the islands in the event of attack. However, they generally followed their own plans, and were primarily interested in carrying out raids on the mainland.[15]

The ROK Marine detachments and the guerillas could not have held these islands without the aid of UN naval forces. Task Element 95.12 cruisers, destroyers, frigates, and ROKN craft, and carrier aircraft of Task Element 95.11 used their fire power to discourage the enemy from concentrating in force on the mainland near Cho-do and Sok-to. Also, their nearby continuous presence presented a strong deterrent to any enemy attempt to mount large-scale amphibious landings.[16]

To the enemy's advantage, hydrographic conditions were not favorable for ships carrying out inshore patrols. The navigable channels were restricted in size; mud and sand banks, and rocks and shoals were in abundance; and strong currents added to the prevailing hazards.[17]

The enemy added to these natural perils through employment of field guns sited along inshore routes used by UN ships, and mines emplaced in the waters through which they passed. From mid-1951, their batteries of field guns were manned by well-trained gunners, amply supplied with ammunition. The mining danger was minimized, however, because the close blockade by naval vessels made even restrictive waters very unhealthy for minelayers. Nevertheless, minesweepers (usually ROK but sometimes USN) were continually at work, check-sweeping the more important navigable channels.[18]

GUERILLA RAID LAUNCHED FROM CHO-DO

Photo 15-2

Canadian destroyer HMCS *Cayuga* (DDE218) during a replenishment at sea with the Australian light carrier HMAS *Sydney* (R17), probably in Korean waters.
Australian War Memorial photograph P05890.055

The Cho-do–Sok-to area was to be a center of activity for Canadian destroyers, from late-summer 1951 through the remainder of the war. On 8 September 1951, while on an island patrol, the Canadian destroyer HMCS *Cayuga* provided gunfire support for a guerilla raid launched from Cho-do. In early morning darkness, the landing force (some 200 men of a guerilla "regiment") embarked in junks at Cho-do. Two motor fishing vessels towed the craft across the strait to the mainland, and the men went ashore on beaches south of the mouth of the Namch'on River. Meeting no opposition, they established a command post from which *Cayuga*'s gunfire could be directed, then moved inland in groups of 30-40 men. Their objectives were to inflict damage on the enemy; gather intelligence for Salamander; and, importantly, secure food for the inhabitants on Cho-do.[19]

The raiders had been ashore for over four hours when, at 0715, *Cayuga* received the first call for naval gunfire support. Subsequent calls were more frequent and insistent, as the enemy pressed the retiring guerillas closely. Requests for fire were passed from the command post ashore to "Salamander" staff aboard *Cayuga*, which relayed target information to her gunnery department. At the height of the action, the destroyer's guns were pouring out rounds as rapidly as their loaders (sailors) could ram them into the breeches.[20]

The last of the guerillas cleared the beaches by 1425, and *Cayuga* departed on patrol. When the destroyer visited Cho-do a few days later, conservative estimates set the enemy losses in the raid at 150 killed and 100 wounded. The guerilla casualties were seven wounded, three seriously. The latter men, one with a rifle bullet embedded in his lung, were treated by *Cayuga*'s medical officer, Surgeon-Lieutenant Joseph C. Cyr, RCN, before being transferred to the cruiser HMS *Belfast*.[21]

GREATEST HOAX IN CANADIAN NAVAL HISTORY

The following month, *Cayuga*'s officers and men would learn to their astonishment, that their esteemed Navy doctor was a fake. The real identity of Lieutenant Cyr was Ferdinand Waldo Demara Jr., an American, accomplished impersonator, and conman. He would become known as "The Great Imposter," and be the subject of a 1961 film by that title staring the American actor Tony Curtis as Demara.[22]

Demara was born in Lawrence, Massachusetts, on 21 December 1921. The many impersonations and scams associated with his adult life are too numerous to describe here, but among them was previous naval and medical experience. He had previously joined the U.S. Navy and, when things did not work out as planned, faked his suicide. After other false identities and occupations, and some prison time, Demara impersonated Brother John Payne and joined the Brothers of Christian Instruction in the State of Maine, a Roman Catholic order. While serving in this capacity, he became friendly with a doctor, Joseph C. Cyr, often visited his office, and presumably picked up some knowledge.[23]

In March 1951, Dr. Cyr walked into the Navy recruiting office in Saint John, New Brunswick, a seaport on the Canadian Atlantic coast, and offered his services. Qualified medical officers were desperately needed at this point in the Korean War by the Army, RCAF, and RCN. Not wanting to lose this potential recruit to a sister service, Cyr's credentials were quickly accepted without verification. Three days later, he was commissioned into the RCN as a surgeon-lieutenant, completing an enlistment process that would normally take two months.[24]

Following a short stint of duty at HMCS Stadacona, an RCN shore establishment in the Halifax area, Lieutenant Cyr joined HMCS *Cayuga* at Esquimalt, British Columbia, on 15 June 1951. She departed three days later, for her second tour of duty in Korean waters. In September, after the casualties of the previously described raid were brought into the ship's operating room, and were being prepped for surgery by his assistant, Demara disappeared into his cabin with a book on general surgery. He speed-read the procedures that he would be required to perform, including major chest surgery. Miraculously, none of his patients died—but then his luck ran out. An account of the removal of a bullet from one of the Koreans, ended up in Canadian newspapers.[25]

One of the readers was the mother of the real Dr. Joseph Cyr. Astonished to learn that her son had apparently performed emergency surgery aboard a Canadian destroyer off the coast of Korea, she contacted him. Flowing reassurance that he was still practicing civilian medicine in Grand Falls, New Brunswick, she contacted the Royal Canadian Mounted Police. In October, *Cayuga*'s commanding officer received a message stating in effect that his medical officer was an unqualified imposter. He found this hard to accept, believing Cyr to be a capable and popular doctor. A follow-on message the next day, dispelled all doubts, and Dr. Cyr was transferred to the cruiser HMS *Ceylon*, for further transport to Japan and return to Canada.[26]

On arrival in Canada, Demara went before a naval board of inquiry which, it appears, wanted to put the whole matter behind them. There is no record of disciplinary action taken, and his service record indicates that Cyr was given an honorable discharge and several hundred dollars in back pay (and quickly deported). Demara later returned to his old tricks, but the publicity he received made subsequent impersonations harder to avoid detection. He died, on 7 June 1982, at the age of 60.[27]

16

Enemy Island Campaign

In autumn 1951, UN forces occupied most of the west coast islands. In support of them, blockade commanders supplied additional patrols and occasional bombardments for Leopard operations discouraging threats to friendly islands. To the Communists, the various activities conducted from these islands were an increasing deterrent to their combat activities, while frequent shore bombardments of the Ongjin and Chulsan peninsulas by naval forces were inflicting heavy damage on their troops and equipment.[1]

However, possessing only limited numbers of craft (mostly sampans and junks used for minelaying), the enemy response to UN Forces activities was also correspondingly light. Two enemy raids were carried out against the islands of Yongmae-do (off the Ongjin peninsula) and Wolto (Cholsan peninsula), in April and September 1951, respectively. Both were repelled by the friendly occupying guerilla forces and by blockade ships. Enemy aircraft activities were also limited, owing to UN air superiority. Although unidentified aircraft had approached friendly islands at night and occasionally dropped a few bombs, the resulting damage was negligible.[2]

The Communists—recognizing that local forces had inadequate trained troops and equipment for amphibious operations necessary to drive UN Forces from the islands—took corrective actions to rectify the situation. These began with putting Chai Zeng Guo, vice commander of the CCF (Chinese Communist Forces) 50th Army, in charge of the island offensive. In mid-October, the 50th began the requisition of equipment and training of troops to prepare for offensive operations against the friendly islands.[3]

When ready, the Chinese began amphibious landing drills, as part of a planned "from near island to far, one by one" joint operation with their Air Force. Concurrently, from 2 November 1951, Soviet-built Lavochkin LA-11 and Mikoyan-Gurevich MiG-15 fighter aircraft conducted air reconnaissance over the islands around the Yalu Gulf area.[4]

TASK GROUP 95.1 UNPREPARED

Toward the end of October, reports were received of Chinese troop infiltration into the Hwanghae area and the northern area around the Yalu Gulf, while ships conducting shore bombardment began experiencing stronger reactions from enemy batteries. Rear Admiral Scott-Moncrieff did not respond to the beginning stages of what turned out to be an enemy planned offensive against the northern islands, because of a lack of reliable intelligence and a shortage of available Task Group 95.1 ships.[5]

At the time, he had restricted his ships' entrance into the Yalu Gulf area, where there was a high risk of mining by sampans along the narrow channels. More importantly, he believed that devoting resources to attempt to permanently hold the islands in the northern area would be more of a liability than an advantage, because the Communists could easily reinforce themselves over the mud-flats. Task group emphasis, then, was on island protection centered on patrolling in close proximity to several important islands south of Cho-do.[6]

CHINESE FORCES LAUNCH ISLAND OFFENSIVE

Photo 16-1

Royal Canadian Navy destroyer HMCS *Athabaskan* during the Korea War. Australian War Memorial photograph P05890.060

On 6 November, the cruiser HMS *Belfast* received a series of messages from Leopard headquarters, reporting that Communist landings had occurred at Ka-do and Tan-do off the west coast of Korea. Subsequently, a report came of an attack by several bombers on Taewha-do in the Yalu Gulf. Under the escort of sixteen LA-11s, nine Tupolev TU-2 aircraft bombed the latter island to consolidate the security of the two recaptured islands by inflicting damage on the guerillas on nearby Taewha-do. This was the first confirmed use of Soviet-built light bombers by Chinese communists.[7]

The commanding officer of *Belfast*, Capt. Aubrey St. Clair-Ford (at that time CTE 95.12), ordered the destroyer HMCS *Athabaskan*, which was already en route to the Yalu Gulf, to evacuate the Leopard representatives and wounded guerillas on Taewha-do. *Athabaskan* had earlier received from *Belfast*, seaward of Sok-to, a quantity of arms and other necessities for delivery to the garrison at Taewha. The destroyer was only 15 miles from her destination when *Belfast* radioed her that Taewha was under attack. As she drew nearer, the island came into view, outlined in the glow of fires burning on the northern slopes, while there came the sound of small-arms fire from up in the hills.[8]

Athabaskan anchored less than two miles off Taewha lighthouse, on the northwest tip of the island, and the local Leopard agent, Lt. A. N. Allan, USA, came aboard. He reported that the bombers had practically wiped out the village on the island, destroyed an ammunition dump, and set afire the main food storage area. Civilian casualties were high, but there had been no attempted invasion by enemy forces. The small-arms fire heard by those aboard the ship were thought to result from jittery guerilla detachments firing at one another.[9]

Lieutenant Allan transferred the supplies *Athabaskan* had brought for him to his boat, and returned ashore, whereupon the destroyer moved northward to defend the northern approaches to the island, should the enemy attempt invasion. That night, making good use of a list of targets on Ka-to, provided by Allan, *Athabaskan* worked them over with 4-inch gunfire. Meanwhile *Belfast*, which had proceeded northward at high speed, anchored seven miles southwest of the island to lend assistance with her 6-inch guns should this become necessary.[10]

There was some delay in the evacuation of casualties, because they had been moved inland and had to be recovered. *Athabaskan* received the 47 wounded and a South Korean nurse who accompanied them, from a junk that came alongside, in early morning darkness, on 8 November. Shortly before 0400, *Athabaskan* and HMCS *Cayuga*, which had provided her cover during the operation, set course for Paengyong-do. Once clear of the danger area, *Cayuga* returned to her previous duty, one of several ships screening the light carrier HMAS *Sydney* (R17), from which she had been borrowed.[11]

Commander, Task Element 95.12, decided that each night at least one ship, and when possible two, should be assigned to the defense of Taewha-do. In *Athabaskan*'s absence, *Cayuga* was reassigned from her carrier screening to perform this duty the night of 8 November. Rather than remaining passively on station watching for invasion craft, Comdr. James Plomer, RCN, took his destroyer northward with as much stealth as possible. At 0345, after closing to within 1,200 yards of Tan-do and Ka-do, and after laying quietly for several hours, *Cayuga*'s guns opened fire, targeting troop concentrations identified by the guerillas on Taewha and deliverd 268 rounds of 4-inch gunfire.[12]

Cayuga returned to Task Element 95.11 at 0952 on 9 November.[14]

Plomer had considerable combat experience. He commanded the Royal Navy corvette HMS *Sunflower* in World War II, and received the Distinguished Service Cross for "gallantry and outstanding service in action with enemy submarines." *Sunflower* sank *U-638* and *U-631* with depth charges, on 5 May 1943 and 17 October 1943, respectively. For his service in Korea in 1951, he was awarded a bar to his DSC (signifying a second award of the medal).[13]

Photo 16-2

Royal Navy corvette HMS *Sunflower* (K41) location and date unknown.
Imperial War Museums photograph FL 4520

ENEMY INVASION OF TAEWHA-DO

On the evening of 30 November, HMCS *Athabaskan* intercepted a "Flash" message from the British destroyer HMS *Cockade* to CTE 95.12, reporting that Taewha-do was being invaded, while she was proceeding from her carrier screening duty to the Bugatti patrol area near the 39th parallel, to act as the early radar warning picket ship. Taking immediate action, Comdr. Dudley G. King, RCN, her commanding officer, anticipating the orders he knew would arrive, increased speed to 30 knots and set a direct course for Taewha.[15]

Earlier in November, *Athabaskan* had turned over command of the Cho-do–Sok-to patrol to the British frigate HMS *Whitesand Bay*, but a different British ship, the destroyer HMS *Cockade*, was assigned the Taewha patrol when the invasion began. Unfortunately, she was not equipped with the superior HDWS Sperry surface-search radar fitted in the Canadian destroyers, and did not detect the invasion force of junks and rubber boats until the first waves reached the beaches. (These type craft would have been difficult to detect in any case, because of their small silhouettes and type of material used in construction. Wood and rubber absorb more energy from radar waves than does steel, for example, leaving less remaining energy to reflect back, and "paint" a contact on the searching ship's radar screen.)[16]

Photo 16-3

British destroyer HMS *Cockade* (D34) in Korean waters, 9 June 1951.
Australian War Memorial photograph P00320.009

Once alerted, *Cockade* opened fire, sinking one junk, probably a second, and damaging several. Enemy batteries ashore then found the range and repeatedly straddled the destroyer with gunfire, scoring one hit. A direct strike on Y-gun killed Able Seaman Clifford Skelton. (He was buried at sea the following day with full naval honors.) Although the destroyer had suffered only superficial damage, her commanding officer could not risk his ship being disabled among the sand banks off Taewha. If he encountered that eventuality at day break, she would have been a stationary target for shore battery fire, and attack by Chinese aircraft.[17]

Unfortunately, the enemy force had gotten ashore, and apparently, had quickly overrun the defenses for, early in the action, the garrison's radio went off the air. *Cockade*, under hostile fire, and unable to provide any meaningful assistance to the friendly forces on the island, prudently withdrew to the south to await reinforcements.[18]

HMCS *Athabaskan* sighted HMS *Cockade* at 0425 on 1 December, and as per orders of HMS *Ceylon*, the two ships remained below the 39th parallel, awaiting air cover to be provided by the U.S. 5th Air Force before taking further action. At 0650, *Ceylon*, which had replaced *Belfast* as CTE 95.12, arrived under air cover from the light carrier HMAS *Sydney*. A few moments later, Air Force jets appeared and took up their CAP (combat air patrol) duties.[19]

Athabaskan, under orders from *Ceylon* to, "Make a quick dash north. Do not linger, but if you can collect any useful evacuees, do so," set a course for Taewha. The Air Force CAP accompanied her, and remained overhead during the impending operation to protect against air attack. During passage to the island, *Athabaskan* sighted junks carrying guerillas and their dependents, making their way south to safety. Arriving off Taewha's eastern shore, the destroyer anchored in the lee of the farther of four nearby small islets.[20]

The scene visible on Taewha was one of disaster and desperation. Fires burned everywhere, and the crackle of small arms fire suggested that the surviving members of the besieged garrison were making a final stand. Small bands of enemy troops could be seen searching the island, apparently for UN personnel or guerillas trapped there. A group having the temerity to direct machine gun fire at *Athabaskan*, received four rounds of high explosive in return.[21]

After sighting two persons on nearby islets (one on Taejongjok-to and the other, on Sojongjok-to) waving frantically, apparently calling for help, *Athabaskan*'s captain, Commander King, decided to shift nearer the latter islet. Its sole occupant had a small skiff near him on the beach. Moving closer to Sojongjok-to (and thus also to Taewha), enemy shells began falling 700-1,000 yards from the destroyer. Just after the grateful refugee boarded the ship, the Communist gunner found the range and straddled *Athabaskan*. (Unfortunately, the individual on Taejongjok-to could not be similarly helped.)[22]

King cleared the area, and arrived at Paengyong-do that afternoon. Meeting with Captain Lloyd-Davies (*Ceylon*'s commanding officer, and CTE 95.12), he received permission to take *Athabaskan* back to Taewha that night to search for other survivors under the cloak of darkness. Just after midnight, on 2 December, *Athabaskan* anchored one mile southeast of Sojongjok-to, and sent in a cutter, carrying an armed party of six led by Lt. Comdr. Charles A. Hamer, RCN(R).[23]

The cutter cruised along the Taewha shoreline for nearly two hours, its occupants shouting loudly at intervals to attract attention. There was, however, no response of any kind, and the boat was recalled at 0200. *Athabaskan* then weighed and set course to rejoin TE 95.11 (the escort carrier USS *Rendova* and her screen of ships).[24]

HIGH COMMAND UPSET BY LOSS OF TAEWHA-DO

The capture of Taewha by enemy forces caused a furor at the UN headquarters in Tokyo and at Sasebo. The UN-held islands north of the 38th parallel were one of the important items under discussion at the peace talks at Panmunjom, and the more of them that fell to the Communists the weaker would be the bargaining position of the UN delegates. In early December, on direct orders from Tokyo, island defense on the west coast was given the highest priority, taking precedence over all other missions, including blockade. Ships on island defense were also given first call on the services of the planes of TE 95.11.[25]

Photo 16-4

UN delegates at Panmunjom, 30 November 1951. From left to right: Maj. Gen. Howard Turner, USAF; Major General Lee, Republic of Korea Army; Vice Adm. C. Turner Joy, USN, Senior Delegate; Rear Adm. Ruthven E. Libby, USN; Maj. Gen. Henry I. Hodes, USA; and Rear Adm. Arleigh A. Burke, USN. Libby was Burke's relief as a delegate. National Archives photograph 80-G-436371

STRENGTHENING OF ISLAND DEFENSES

Several steps were taken to improve the defenses of threatened islands. Mortars and machine guns, and later Bofors and Oerlikons, were brought in by UN naval forces, and positioned to cover the beaches at which enemy landings might be made. The next challenge was one of manpower. There were insufficient ROK Marines on the islands, and the guerillas, unused to discipline and familiar only with hit-and-run offensive tactics, were not ideal for garrison duty. The problem of island defense forces was later partially solved by bringing in more ROK regulars, but there were never enough of them.[26]

A new organization was created, with headquarters on Paengyong-do, with relatively strong detachments stationed on all the major islands. (The Leopold raiding and intelligence-gathering organization continued to exist side-by-side with, but subordinate to, the new island defense force.) The improved command and control brought in for the new organization simplified, somewhat, the challenge of coordinating the activities of the naval forces, guerillas, and ROK garrison troops. However, the intelligence organizations on the islands continued to operate independently, and to steadfastly reject any close cooperation that would disclose their current activities. Similarly, there was seldom any very close cooperation on the west coast islands between the navy, marine, and army groups nominally working together, and those of the United States Air Force.[27]

REORGANIZATION OF TASK ELEMENT 95.12

Task Group 95.1, under Rear Adm. Alan K. Scott-Moncrieff, RN, had been organized, since 10 April 1951, as shown below:

Task Group 95.1	
TE 95.11	West coast aircraft carrier and screening ship
TE 95.12	Patrol and blockade on west coast
TE 95.13	Escort frigates, administrative purposes only
TE 95.14	Special gunfire support element – formed as required
TE 95.15	Special small operations element – formed as required
TE 95.16	Sometimes formed when additional minesweeping resources were allotted to TG 95.1. These sweeper forces would usually also retain the element number assigned by their own commander, CTG 95.6[28]

The task element structure remained the same, but the Patrol and Blockade Element (TE 95.12) was then split into four smaller task units to deal more effectively with the Communist threat to the major islands. The Cho-do–Sok-to naval force was re-designated TU 95.12.1, and it normally operated under the command of one of the Royal Navy destroyers or *Bay*-class frigates. The island defense assignments of the four task groups are identified below:

- Task Unit 95.12.1: Cho-do–Sok-to area
- Task Unit 95.12.2: Paengyong-do
- Task Unit 95.12.3: Han River islands
- Task Unit 95.12.4: Yongpyong-do[29]

ENEMY ISLAND OFFENSIVE CONTINUES

Photo 16-5

George C. Dyer, rank at the time and date of photo is unknown.
Naval History and Heritage Command photograph NH 82847

As winter fell, again bringing strong winds, cold, and snow to the northern Yellow Sea, the ships of the island defense forces took up their patrols. On 7 December, Rear Adm. George C. Dyer, commander, Task Force 95, arrived at Cho-do aboard the light cruiser USS *Manchester* (CL-83). His flagship was followed in by HMS *Ceylon*. At Cho-do and

Sok-to, Dyer found morale improved by the news that the islands would be defended, but the situation was still precarious.[30]

Discussion with Leopard officers indicated a great need of getting the refugees out and the ROK Marines in as fast as possible. As a consequence of the meeting, a USN dock landing ship (LSD) and some wooden-hulled minesweepers (AMS) soon arrived to keep the Sok-to anchorage swept and to strengthen the small craft patrol. Arrangements were also made for the LSTs (tank landing ships) bringing up the ROK Marines, to remove the refugees from the islands. With these actions undertaken, and an apparently growing enemy small boat menace to the Wonsan harbor islands, Dyer proceeded back to the east coast.[31]

The admiral had barely reached Wonsan when word was received of attacks on Ung-do and Changyang-do, two small islands shoreward of Sok-to. Despite support from UN ships and aircraft, an enemy force of about 600 overwhelmed the guerillas on the islands in a night attack, on 17 December, and all resistance on them ceased. With the situation still deteriorating, Dyer again headed west.[32]

By 20 December, the growing number of ships on anti-invasion duty near Cho-do included the cruisers USS *Manchester* and HMS *Ceylon*, destroyers USS *Eversole* and HMS *Tobruk*, and the frigate HMS *Alacrity*. On that day, HMS *Tobruk*, while covering LST loading at Cho-do and Sok-to, received one hit from a distant enemy battery, but sustained no damage. Four days earlier, HMS *Constance* had been hit by enemy shore fire from Amgak Peninsula, which did not result in personnel casualties, but rendered her sonar, and degaussing equipment used for protection against magnetic mines, inoperative.[33]

On 22 December, a tank landing ship in the vicinity of Sok-to, hit by enemy shore fire from Amgak, received only superficial damage. Two days later, the group of LSTs of which she was a part, completed the evacuation of 7,196 refugees from Cho-do, Paengyong-do and Taechong-do islands.[34]

ISLANDS CONCEDED, BUT DEFENSE CONTINUES

One day earlier, on 21 December, the UN negotiators conceded all islands north of 38° N. Latitude to North Korea. However, failing an armistice agreement, the defensive requirement of islands remained. On 6 January 1952, responsibility for the overall defense, including local ground defense, of designated islands on both coasts, was assigned the UN naval forces and delegated to commander, Task Force 95.[35]

17

Time Spent "Off the Line"

Immediately on entering the living quarters, we were set upon by a "large-sarge-in-charge," and he hastened to lay down the law. He was quite an amusing character, his vocabulary was limited to "Youse guys," "lissen here," and "I don' wanna catch ya makin' a pest of yourselves." Between babbling about the brig and being sent back to Sasebo, and all sorts of horrible threats, the future was beginning to look dim. The sarge turned out a pretty right guy, tho'.

—From a story in HMCS *Athabaskan*'s newsletter, Christmas edition December 1951, about some of her sailors travelling from Sasebo to Camp Wood, a U.S. Army installation on the island of Kyushu, to enjoy some R&R following combat duty in Korean waters.[1]

Photo 17-1

Painting by Frank Norton of HMAS *Warramunga* at wharf in Kure, Japan, 1952. Australian War Memorial photograph ART40031

Between periods of combat duty in Korean waters, UN naval force ships typically entered southern Japanese ports: to receive much needed maintenance; for resupply of provisions, stores, and ammunition; and to allow their crews some rest and relaxation ashore. These ports were generally restricted to Kure or Sasebo. Kure was regarded as the base for British and Commonwealth ships, whereas Sasebo was used mostly by American ships. However, ships from several countries used both ports, including those of the Royal Navy, Japan being more convenient for them than visiting their more distant bases at Hong Kong and Singapore.[2]

The Royal Australian Navy base HMAS Commonwealth, at Kure, was the main arrival point for RAN warships bound for service in Korean waters, and the primary logistics support facility in Japan for the Royal Australian Navy. Kure was also a popular liberty port and the main Commonwealth Forces hospital was also located there.[3]

For the U.S. Navy, the port of Sasebo in western Kyushu, only some 165 nautical miles from Korea's southern end, was a nearly ideal base location. Though almost 700 miles by sea from the combat area, Yokosuka (located near Tokyo) offered a superb maintenance facility, and was much closer than any other comparable naval shipyard for the Allies.[4]

Photo 17-2

Repair ship USS *Ajax* (AR-6) at Sasebo, Japan, 14 December 1952. Ships nested along her port side include (left to right): USS *The Sullivans* (DD-537); USS *McGowan* (DD-678); USS *Lewis Hancock* (DD-675) and ROKN frigate *Imchin* (F-66).
National Archives photograph #80-G-478507

The UN forces were fortunate to have Japan as a forward base with its ports, airfields and other facilities, and hardworking and capable citizens. The Japanese people were a priceless asset in keeping the operating forces well-supplied and repaired, and in providing relaxation and recreation opportunities for service members.[5]

CAMP WOOD "THE REST CAMP"

The following account typifies how UN sailors received R & R during the Korean War. In December 1951, the Canadian HMCS *Athabaskan* was at Sasebo for the Christmas holiday. One morning, some forty members of her crew set off for Camp Wood (a U.S. Army installation), for rest and relaxation, "Piped awake" at 0530 on a cold, grey dawn, the sailors packed their bags, departed the ship, and were on their way.[6]

Sasebo is a seaport on the west coast of Kyushu, the most southerly and scenic of the four Japanese home islands. Forewarned by shipmates, who had previously made the long trek to the Army camp (located near Kumamoto in Central Kyushu), they managed to acquire a case of beer to help smooth out the long train ride ahead. After piling into a train car, they settled in. In about an hour, the noise in the car was deafening, as every "Tar" (sailor) on the train was making like a potential Mario Lanza (an Italian tenor), and singing lusty sea chanties which likely would have included the obscene *North Atlantic Squadron* as well as some versions of *Heart of Oak*.[7]

For illustration purposes, the lyrics of *Heart of Oak* are presented below. The original version was composed for the stage by William Boyce, and words written by David Garrick, in 1759. *Heart of Oak* is the official march of the Royal Navy and of several Commonwealth navies, including the Royal Canadian Navy and the Royal New Zealand Navy. It was the official march of the Royal Australian Navy, but has been replaced by a new march, *Royal Australian Navy*.

Heart of Oak
Come, cheer up, my lads, 'tis to glory we steer,
To add something more to this wonderful year;
To honour we call you, as freemen not slaves,
For who are so free as the sons of the waves?

Chorus:
Heart of Oak are our ships,
Jolly Tars are our men,
We always are ready: Steady, boys, Steady!
We'll fight and we'll conquer again and again.
We ne'er see our foes but we wish them to stay,

They never see us but they wish us away;
If they run, why we follow, and run them ashore,
For if they won't fight us, what can we do more?

They say they'll invade us, these terrible foes,
They frighten our women, our children, our beaus,
But if they in their flat-bottoms, in darkness set oar,
Still Britons they'll find to receive them on shore.

We still make them fear and we still make them flee,
And drub them ashore as we drub them at sea,
Then cheer up me lads with one heart let us sing,
Our soldiers and sailors, our statesmen and king.[8]

Time flew on the wings of music and, in no time, the train was approaching the station at Kunimoto. From there, the sailors were herded into two large American Transport buses and taken to Camp Wood, about five miles north of the Kumamoto city center. Encompassing 153 acres, the camp comprised a small city in itself, one that offered many forms of recreation and entertainment.[9]

Directly in front of the Command Post was the Regimental Parade Ground, which also doubled as a football field and the best baseball diamond on Kyushu during appropriate seasons. Among the many recreational facilities were a snack bar, where a cold coke and sundae were in order after a show or a hot day in the field; a tavern; a bowling alley with six lanes; 4/5/6 Club; Non-Commissioned Officers' Club, Officers' Club; and Diamond Theater. There was also within the dependent housing area for officers and NCO's, a golf course, skeet range, NCO Club, and area THREE (a training center) about three miles northeast of Camp Wood.[10]

After entering their living quarters, and being greeting by the "large-sarge-in-charge" referenced earlier, the sailors obtained their blankets, pillows, and sheets. They all wanted a nap before proceeding to anticipated activities. This was not to be, as detailed by the unknown author of an account of this trip published in the ship's newspaper:

> We drew our blankets, pillows, and sheets, from the stores and immediately upon making up our beds we all pulled a T.U. routine [took a nap]. The effects of the train ride had caught up with us. About 20 minutes after crashing, a great commotion was heard; by the way personnel were activated I decided to investigate, and I wish I hadn't. It seems that there was an exercise air-raid and before we could get our wits about us, we were all herded into a bean-bag sized shelter. There we stayed for the next hour, and when the ordeal was

over, all hopes of sleep had vanished, so we trudged to the mess hall for a supper of .... b-e-a-n-s....![11]

The evenings activities centered around a Christmas Party, consisting of a bingo game, and then some wets (alcoholic beverages). The sailor who recorded this account, went to the bingo game for a while, and found they were auctioning off 3-day passes. He didn't try to win one, though, explaining:

> I thought of what the 'Jimmy' [second in command of HMCS *Athabaskan*] would say if I came back to the ship 3 days adrift, and told him I had bought a pass for seven dollars. I think he would have taken a dim view of it, to say the least.

> Now to the Wets and then to bed. There is nothing more comforting in this world than going to bed and knowing that you don't have to get up in the morning. Most of us were cursed with habit tho', and eight-thirty the next morning we were all on our way towards the Service-club to get a cup of coffee to start us on our restful way.[12]

Golfing, basketball, billiards, bowling, and other sports, all reputed essences of restfulness and relaxation, were crammed into a short day; with the only casualties suffered being stiffness of muscles and aching bones. Suppertime brought messmates using both hands to shovel food into their ravenous mouths; but some were so tired, the only way they could eat was wait for a yawn, and then cram. In most cases, a short half-hour nap and shower were sufficient revival to make it to the cocktail bar for a small shot of inspiration and strength. At least one of the drinks was memorable, but not in a good way:

> In case any of you 'old tars' think the only drink in the world has a rum base, I recommend that you try a Camp Wood Whisky Sour.... guaranteed to take the floss off the dirtiest tongue.[13]

That night an Air Force Glee Club visited the camp. Following their performance, they joined the sailors in the cocktail bar. Since they were all singers, and by this time, the sailors were also all "singers," the evening ended with a sing-song. The time spent with the airmen included some kidding at their expense:

> Gad, those 'trappers' look posh in their uniforms; if we had similar suits the percentage of re-engagements would probably increase. To tell the truth, on first sighting the troupe, Lonvik mistook them

for a traveling convention of Grey Hound bus drivers and was heard to ask the time of the next bus to Calgary. Evidently, he wanted to get to the Stampede.[14]

After this joyful fellowship, and ready to fall into their beds for some much anticipated and needed sleep, the sailors were reminded yet again that, while they were not in Korea, they were not free of activities spawned by the war:

Later we returned to the block, feeling mellow, only to be informed that Camp Wood was in condition "Flash Yellow" and were 'told-off' to an air-raid shelter. This met with loud groans and moans, as we were all tired from a hard day getting rested. Fortunately, the senior 'sober' hands quieted the argumentative lads and so into the shelter. But at last the alert was over, and into....ahh....BED!!![15]

Following interludes at Sasebo, Kure, or nearby sites for R&R, and having armed and provisioned, UN naval forces ships stood out to sea, for return to Korean waters, and resumption of war duty.

Photo 17-3

Drawing by Frank Norton of the Australian frigate HMAS *Condamine* passing through Shimonoseki Strait, between the Kyuku Islands, Japan, on her way to Korea, 1952. Australian War Memorial photograph ART40048

18

Tom Tiddler's Ground

These islands in the South are therefore a kind of Tom Tiddler's Ground, where either side can occupy them at will, but neither one is prepared to expend a great effort to hold them. Only if our main islands are threatened do we allow the enemy to tie us down.

—Rear Adm. Alan K. Scott-Moncrieff, RN, explaining the practice of Communist forces to periodically occupy small inshore islands off the south and west coasts of the Ongjin peninsula, remain on them for only one or two days, and then return to the mainland.[1]

Photo 18-1

Coating of ice formed on the Australian frigate HMAS *Condamine* (F698), circa 1952. Australian War Memorial photograph 305195

On 1 January 1952, with the approximate position of the battle line in Korea extending from a point about five miles SE of Kosong on the east coast, westward to a point ten miles SE of Kaesong on the west, the outlook regarding negotiations in progress was dismal. The hopes of an armistice raised by agreements reached at Panmunjom, in November and early December 1951, were fading rapidly with the two sides in disagreement over the prisoner-of-war issue. Some progress had been made at the beginning when both sides exchanged lists of prisoners held, but it soon became clear that after that, any headway would end. Up to this point in the war, United Nations' forces had captured tens of thousands of Communist soldiers. Many of those claimed they had been coerced into fighting for China and North Korea, and did not want to return to their home countries. This presented a serious stumbling block to peace negotiations. North Korea and China insisted that their respective POWs be repatriated, and the United States and South Korea refused to do so on humanitarian grounds.[2]

As the Communists' attitude hardened, it became obvious that no amount of haggling would settle the prisoner issue unless one side or the other could be forced, by other than diplomatic means, to make major concessions. The difficulty for the UN Command was that the military position was no longer what it had been in June 1951. During the months of static warfare, the Communists had not been idle. They had, instead, strengthened their defenses enormously. There was some increase in troop strength, but the main emphasis had been on increasing fire power. Despite UN naval and air forces strikes, the Communists had developed a large and effective force of shore guns and mortars, and a more than adequate stockpile of ammunition.[3]

When it became obvious to the UN negotiators that their reliance on the sincerity of the Communists had been misguided, it was too late to resume the offensive. Any attempt to make an all-out, frontal attack, on the now-fortified enemy positions, would have been far too expensive in men and materiel and, moreover, it was doubtful whether the UN forces possessed the necessary manpower to do so. The UN Command was caught in a trap, partially of its own making, and then had no viable alternative to continuing the stalemated ground campaign, other than to hope for some progress in the discussions at Panmunjom.[4]

However, no matter how the truce talks were progressing or what the situation was on land, the UN ships in Korean waters were kept busy. Their duties included blockading the enemy coast, preventing seaborne attack on friendly islands, harassing enemy coastal supply lines, and providing gunfire support to the troops holding the seaward flanks of the UN front line. During interludes from these ongoing imperatives, the ships engaged in shelling practically any target whose destruction would hurt the enemy's war effort.[5]

HARD, SPARE CONDITIONS AT SEA

Before progressing further with this account of naval operations off the west coast of Korea, in winter 1952, some perspective of life aboard ship, aside from that associated with combat operations, is in order.

The day to day life of sailors aboard any Navy vessel operating at sea, centers around continual work and watchstanding requirements, while enduring sleep deprivation, and living together in cramped quarters. Decades after the Korean War, former HMS *Concord* (D63) crewmember, Derek Hodgson, described his duty aboard the British destroyer. In Royal Navy parlance, "rating" refers to an enlisted sailor, and "mess" to one of several living/eating spaces in which groups of sailors resided aboard ship. Hodgson also described the challenges associated with making oneself presentable to go ashore, when *Concord* was off war patrol. The following account closes with him expressing pride in having served in the Royal Navy:

> Life on board a destroyer during the fifties, was not very different to that experienced since the beginning of the century. An open bridge which during the Atlantic, Arctic and Korean wars where almost arctic conditions were experienced during the winter, was a terrible experience for those on watch for hours on end.
>
> The ratings mess, was a highly suitable name, many men were crowded together in an unimaginable confined space where they were expected to eat, sleep and spend their time off watch. The men took turns to be "cook of the shack" when they were expected to conjure up a meal twice a day, when stores after a few days at sea were in short supply.
>
> In the early commissions, fresh food ran out after four days and a lot of imagination had to be used to produce a decent meal out of tinned, and disgusting dehydrated ingredients…. Fresh water was in short supply and showers used salt water. No one used them, as the stink of stale sea water smelt like rotten eggs, so that meant bathing either from a hand basin or a bucket, as for trying to shave before going ashore, many a bleeding face resulted.
>
> Imagine the scene when preparing to go ashore trying to look "tidy," someone trying to do a late bit of ironing, others frantically looking for clothing stored in seat lockers which usually contained loads of cockroaches, as well as the item of clothing being searched

for. A London Tube [crammed underground electric train car] is a luxury compared to a mess deck preparing for liberty. But you know what! We would not have missed it for the world such was the comradeship, the caring for others, the humour, honesty and being part of a service, which instilled a pride and spirit almost unknown in any other walk of life.[6]

Photo 18-2

Vice Adm. Rollo Mainguy, RCN, chief of Naval Staff, seated with crew members in their mess on board the Canadian destroyer HMCS *Athabaskan* (DDE219). Library and Archives Canada photograph AN-210

NAVAL FORCES' DEFENSE OF SOUTHERN ISLANDS

While UN negotiators had conceded all islands north of the 38th parallel to North Korea, the UN Command was committed to retaining possession of several offshore islands just south of the demarcation line, and retaining control of strategic ones north of the line, until ratification. China was equally committed to capturing them, to rule out their use as a possible bargaining chip for the UNC in the negotiations. Accordingly, Communist forces assaults against islands in the area south of the line continued. However, conducting offensive operations against UN forces-held islands off southern Hwanghae Province was not as easy as in the Yalu Gulf to the north.[7]

Communist forces which were relatively strong in the Cho-do-Sok-to area, had easy access to these northern islands. Thus, Allied defense of the islands was challenging. Conversely, to the south, UN Forces maintained well-prepared defense positions on vital islands. This condition made it difficult for the Communists to conduct successful direct offensives on them. Thus, rather than attempting direct attacks on major islands, as part of their policy "from near island to far, one by one," they pursued capturing inshore small islands first, which had few defensive forces.[8]

Map 18-1

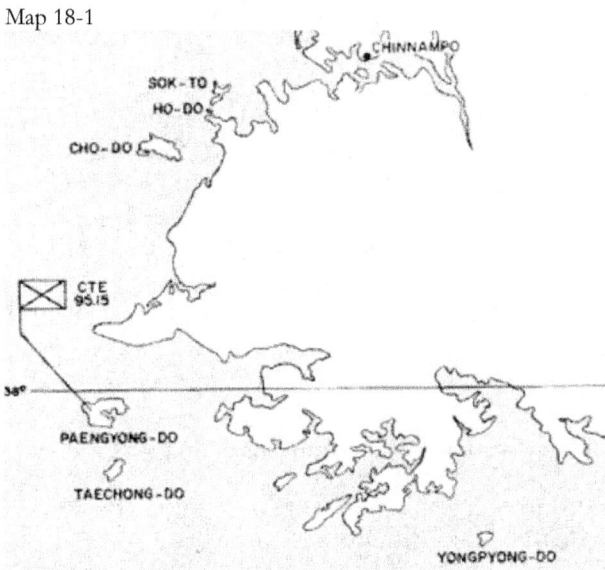

Hwanghae Province and major offshore islands

It was UN policy to not allow the Communists to permanently hold any of the islands along the coast of Hwanghae Province. Most were of no tactical importance, but it was feared that enemy forces might use them as stepping stones, to launch attacks on one or more of the six islands, which the UN Command had designated strategically vital. These included Sok-to and Cho-do north of the line, and the other four listed below, which were south of the line:
- Sok-to (no important facilities, stepping stone to Cho-do)
- Cho-do (site of UN radar stations and SAR personnel)
- Paengyong-do (Leopard headquarters)
- Yongpyong-do (site of Air Force beacon unit)
- Taechong-do (Air Force Shoran parties and Radar stations)
- Tokchok-do (Air Force Shoran parties and Radar stations)[9]

This did not mean that the UN Forces were indifferent to patrolling and controlling the other islands. In addition to potential stepping-stone use, enemy occupation of minor islands near the mainland could threaten UN minesweeping activity in inshore areas, and also prohibit their use by friendly guerillas for launching raids and intelligence-gathering activities ashore.[10]

Thus, the competing interests of the combatants began a new round of island warfare for small and less important islands lying between the major islands and the mainland, which Rear Admiral Scott-Moncrieff characterized as "Tom Tiddler's Ground." This reference was to a British children's game in which a player designated Tom Tiddler, tries to catch the other players who invade the area designated as his/her property. Such an area is often a kind of "no-man's-land" where pickings may be sought or had without effective interference, as players dart in, remain a short time, then flee the area to avoid capture.[11]

Realizing that the defense of the islands would be a prolonged high-priority commitment, Scott-Moncrieff organized the ships of Task Element 95.12 into four task units. Under the command of CTE 95.2, each task unit was assigned a patrol area, identified by a unique code name, as indicated in the table.

Task Element 95.12 Patrol Areas

Task Unit	Title	Patrol Code Name and Area
TU 95.12.1	Sok-to-Cho-do Unit	CIGARRET: Sok-to to Choppeki Point
TU 95.12.2	Peongyong-do Unit	WORTHINGTON: From Choppeki Point, eastward to approximately 125°15'E, including Wollae-do, Yuk-to and Kirin-do
TU 95.12.3	Han Unit	GUINNESS: Han Estuary, but frigates not to proceed beyond Fork anchorage
TU 95.12.4	Haeju Unit	BRICKWOOD: from Worthington area, eastward[12]

Task Unit 95.12.4 was assigned the Haeju area (Brickwood patrol), where a meandering shoreline framed coastal waters containing a mass of islands, many of them small. Some nearshore islets (small islands) almost abutted a confused backdrop of peninsulas, so heavily indented, it was difficult to distinguish between island and mainland.[13]

Task Unit 95.12.4 headquarters were on Yongpyong-do; an island at the mouth of the Haeju-man (bay). Surrounding Yongpyong-do were several groups of islets, identified below:

- To the south: five small islets, and a larger one, Soyongong-do
- To the north between Yongpyong-do and the mainland were the islets Sok-to (not to be confused with the island near Cho-do), Kal-to, Changjae-som, Mudo and Yuksom
- Northeastward in Haeju Bay were several more islets, the most important being Taesuap-to, Sosuap-to, and Yongmae-do[14]

Map 18-2

Haeju area, and islands to the west below the 38th parallel

Seventeen miles westward of Yongpyong-do lay Fankochi Point (or Tungsangot), the southernmost tip of a long, deeply indented peninsula. "Around the corner," as it were, from Fankochi (between that point and the Paengyong Group of islands to the northwest), was yet another group of important offshore islands. The defense of this latter group was shared with the Yongpyong task unit (95.12.2).[15]

The largest island of the Paengyong Group was Sunwi-do, which lay only some three-quarters of a mile from the Fankochi Peninsula. From Sunwi, along a line extending roughly northwest, were Ohwa-do, Changnin-do, Kirin-do, and Mahap-do, the latter lay just off the westerly tip of Yungmahap Point and due east of Paengyong-do. Each of these islands was to be the scene of fighting between friendly guerillas and Communist troops and, in the months to come, UN destroyers would become very familiar with the surrounding waters.[16]

ISLAND DEFENSE IN EARLY 1952

Following heavy losses of guerillas killed defending islands, the Leopard Headquarters issued a direction stating, "Except for the vital islands, if there are enemy attacks on the guerilla-held islands, immediately report this information to the Headquarters and Task Unit Commander, and then withdraw without any trial to secure them." Rather than trying to hold less important islands by use of defending guerilla forces, which had little training for such a role, a new policy was enacted. It was to inflict heavy damage on enemy troops through air strikes and naval gunfire as they attempted landings.[17]

Between late December 1951 and late January 1952, UN attacks on enemy captured islands and defensive activities against additional threats were conducted mainly by naval gunfire and air strikes. Each blockade ship's defensive and harassing activities took place at nighttime. During these darkened periods, ships took station between vulnerable islands and the mainland, and illuminated the sky with star shells to help prevent enemy seaborne attacks by junks and smaller craft. Concurrently, shore bombardments were carried out against the enemy-held islands within gunfire range. Areas further inshore were patrolled and harassed by ROKN craft.[18]

By dawn, the ships normally withdrew out of the immediate range of shore batteries and mortars. Daytime attacks were conducted by the aircraft of TE 95.11, at that time, from HMAS *Sydney* and USS *Badoeng Strait*. The guerillas' role was reporting enemy movements on the captured islands and mainland, and recapturing islands with weak or no enemy defenses.[19]

GUNNERY / PATROL DUTY ABOARD HMS *CONCORD*

> *We saw action against North Korea and had several escapades, people got lost and once we nearly went the wrong side of Formosa [Taiwan] – but got away with it! Concord always had spark and good humour, with an aura of confidence and efficiency. Above all we had 'Pride of Ship' which continues to this day.*
>
> —Comdr. Rodney Agar, RN (Retired), former gunnery officer aboard the British destroyer HMS *Concord* during the Korean War.[20]

Gunnery Branch
Always listen to the Gunnery Officer - then do it your way.
Take plenty of time doing "Preps for firing"- You are allowed ten minutes.
Be prepared to make anything - at any time - with no materials.
Always be polite to the G.I. [Gunnery Instructor] - it spoils his whole day.
Always be clean and tidy - cheerful and willing - abstain from drinking and smoking - if you can do all this, you should not be in the gunnery branch.

—Author unknown.[21]

Photo 18-3

British destroyer HMS *Concord* (R63) during World War II, 10 December 1946. Imperial War Museums photograph FL 8326

One of the UN Naval Force ships assigned island defense duties off the west coast of Korea was the British destroyer HMS *Concord*, which did six tours in the theater during the war:
- September 1950 to January 1951
- April to May 1951
- August to December 1951
- January to May 1952
- June to November 1952
- May to July 1953[22]

During the early part of 1952, Lt. Rodney Agar reported aboard the destroyer as her new gunnery officer. Excerpts from letters home to his parents reflect his youthful perspective, and offer insights on *Concord*'s operations.

7 March 1952:
> This ship is undoubtedly jolly good, with a very nice crowd in the wardroom and I'm thoroughly enjoying life! The next two years should be great fun and a great experience. Last night we did a short bombardment of an enemy held village, and boarded a couple of junks. Not very important incidents but my baptism!! It is really rather fascinating creeping in and out between these masses of small islands in narrow channels, all of which are continually being captured and re-captured by the enemy ah-jongs.... The [Communist] guerillas creep about in junks and set up gun positions ashore and it is our job to continually harass them, and prevent them from capturing islands and also bombard them out of the islands they have already captured. During the day we have been escorting one of the American carriers, who flies off strikes from dawn till dusk. We also go to action stations every dawn at 0630, irrespective of what's been going on the night.

18 March 1952:
> We are patrolling a group of islands in the Haeju area on the West Coast, preventing the enemy from capturing them, and also investigating junk traffic. We have a number of Korean ah-jongs on the islands, with some American marines, and they sally forth to make raids inland. The enemy are building up strength on the mainland and have occupied villages and dug trenches so our job is to keep them quiet. We have bombarded every night except one. The first night we fired at two villages, containing troops, and destroyed several houses and saw two big explosions. Later on that night we repulsed an enemy invasion of one of the islands by 4.5 [inch] gun fire and everyone was very pleased, though we were using a very Heath-Robinson method of fire control with me and the proper fire control parties not even turned out!! We were lucky to hit the target really.
>
> The next two nights we fired the odd shell at trenches and villages and bombarded lookout positions with Bofors. We usually fire starshell during the night as it has a good moral effect and lights up the mudflats across which the enemy have to advance. This and keeping watch one in three keeps one very busy and life is not dull!! Yesterday I went ashore with the Captain to look at the main island which we are responsible for, and we met the three U.S. soldiers who are there, one of whom was the garrison commander. It was very interesting and the Americans were very nice. We had them back for lunch and gave them magazines, booze etc, and they had baths on board! Life is rather primitive for them.

28 March 1952
> Well, it is nice to have a rest for a change and we shall be in Kure for maintenance till 8th April when we sail for the East coast and do another patrol of 18 days – rather longer! It is really good fun and thoroughly interesting – last patrol we fired nearly 500 rounds at enemy troops and villages, observation positions etc. with quite good results, so we gather. The weather has been very kind to us except for a few days when it blew, and at the moment we are just entering the Inland Sea in Japan where it is just like an English Summer's day.[23]

LT. AUGUSTUS AGAR VC DSO, ROYAL NAVY

Lt. Rodney Agar had much to be proud of, other than his own Royal Navy and Korean War service. While a RN lieutenant, his grandfather, Augustus Willington Shelton Agar, had earned both the Victoria Cross and Distinguished Service Order in 1919. The qualifying action for the VC, Great Britain's highest award for heroism, was so highly classified, a description of it could not be listed on the award citation, nor was any public announcement made. While this event is pretty far afield from the subject of the book, it is worth devoting a page or so to it.[24]

In 1919, Lt. Augustus Agar was serving at the Royal Navy's Coastal Motor Boat base at Osea—a small island in the estuary of the River Blackwater, Essex, East England, at which torpedo-carrying speedboats were based. Agar was approached by Sir George Mansfield Smith-Cumming, the first head of the foreign section of the British Secret Intelligence Service (MI6), to volunteer for a mission in the Baltic.[25]

The mission was to bring a British agent back to England from Bolshevik Russia. Paul Henry Dukes had infiltrated the Bolshevik government, had in his possession copies of top secret-documents, and was stranded in Petrograd (today St Petersburg). The borders were sealed, and it was time to get him out and away from the Cheka (secret police). So, a rendezvous on the coastline near Petrograd was planned to retrieve Dukes.[26]

To do this Agar's boats, operating from Terijoki, on the north coast of the Gulf of Finland and close to the Russian border, would have to cross Bolshevik minefields; pass the island fortress of Kronstadt; and evade gunboat patrols, submarines, and seaplanes guarding the entrance to the Bolshevik naval base at Kronstadt. After making it this far, it would be necessary to continue on to Petrograd to pick up Dukes.[27]

Before making the rendezvous point, Agar decided to launch an attack at Kronstadt, on the Russian cruiser *Oleg*, which had been hammering the White Russian garrison trapped in the nearby fortress of Krasnaya Gorka. On the night of 17 June 1919, two boats and their three-man crews set out, with the intention to sink the cruiser and her crew of 565. One the boats had to abort with engine failure but, undeterred, Agar continued on alone into the bay. With great skill and seamanship, in Coastal Motor Boat *CMB-4*, he slipped past three destroyers and launched a torpedo at the *Oleg* from a distance of about 900 yards, which sank her. Agar then sped away under heavy fire.[28]

As a result of this raid, Lieutenant Agar was awarded the Victoria Cross. Because the Russians put a price on his head of £5,000, his receipt of it was not openly acknowledged and this Victoria Cross became known as "the mystery VC." Agar's actions also gained him promotion to lieutenant commander, on 30 June 1919.[29]

17 June 1919
Royal Navy Coastal Motor Boat *CMB-4*
sank the Russian cruiser *Oleg*

18 June 1919
six Royal Navy coastal motor boats
damaged two Russian battleships:
pre-dreadnought *Andrei Pervozvanny*
dreadnought *Petropavlovsk*
sank Russian submarine depot ship:
Pamiat Azova

Victoria Cross   **Distinguished Service Order**

Early in the morning of 18 June, Agar went out again, this time with six other CMBs accompanying his boat to the Russian naval base. At a cost of eight personnel killed and nine captured, the group damaged two battleships—the pre-dreadnought *Andrei Pervozvanny* and the dreadnought *Petropavlovsk*, and sank the submarine depot ship *Pamiat Azova*. Two additional Victoria Crosses were awarded for this raid and Agar received the Distinguished Service Order.[30]

Agar had yet to fulfill his orders of rescuing the spy Paul Dukes. Successive attempts to pass the fifteen forts associated with the island fortress of Kronstadt failed. The thirteenth such went disastrously wrong; so much so that the Russian propaganda convinced Dukes that Agar and his men had been killed in the attempt to rescue him. At this point, he abandoned all hope of rescue by sea and left Petrograd by land, jumping from tram to tram to evade the Cheka and escaped through Latvia using a variety of disguises. Dukes, whose nickname was the "Man of a Hundred Faces," made it back to London with the tissue paper on which he had copied the secret documents intact. Dukes was knighted, the only person ever to receive this honor based entirely on their exploits as a spy.[31]

19

Ongoing Island Defense

I shall long remember, the sight of twelve sailing junks, in groups of three to five, each overflowing with troops, being towed towards the enemy coast at a rate of about 21 knots by motor junks powered with asthmatic engines.

—Comdr. Dudley G. King, RCN, commanding officer, HMCS *Athabaskan*.[1]

Photo 19-1

Canadian destroyer HMCS *Athabaskan* (DDE219), circa 1952.
Courtesy of John MacFarlane

During a patrol, from 8-16 June 1952, by the Canadian destroyer HMCS *Athabaskan*, her commanding officer, Comdr. Dudley King, RCN, was assigned as commander, Task Unit 95.12.4. When relieved by the frigate HMS *Amethyst* on the morning of 16 June, *Athabaskan* had fired, in the seven-day period, 1,607 rounds of 4-inch and 2,231 rounds of 40mm. It was an active patrol, with action beginning the first night.[2]

On 8 June, the destroyer was lying at anchor off Taesuap-to, in the upper estuary of Haeju Bay, when tracer fire was observed in the vicinity of Mudo. Realizing that Mudo was being attacked by sea, *Athabaskan* ordered the tank landing ship *LST-1089* to take over her station, and then raced to provide assistance. When she reached Mudo, the attack was over, repulsed by the forty or so guerillas on the island.[3]

The enemy raiding force had been embarked in six sailing and one motor junk. The guerillas claimed to have sunk four of the sail at a cost to themselves of two men wounded. The next morning, a good deal of wreckage was sighted floating in the area. Before departing, *Athabaskan* sprayed the mainland opposite Mu-do with 667 rounds of 40mm to discourage any additional enemy action.[4]

The following night was quiet, but on the 10th, the enemy initiated another island attack. *Athabaskan* was at her night-station east of Mu-do when, at 2200, gunfire was heard from the mainland to the northwest of the island. A few moments later, tracer fire was seen in the direction of Yongmae-do, across the bay off the Ongjin peninsula. Commander King, was faced with a quandary. While suspecting that gunfire associated with one was a feint and the other a real attack, he chose Yongmae-do to lend assistance to, the much more important island.[5]

Athabaskan set course eastward, firing star-shell over the channel separating Mu-do from the mainland as she left, in an effort to mislead the enemy into believing that a UN ship was off Mu-do. With the state of the tide, low water, the destroyer could get no closer to Yongmae than the Taesuap-to area. The guerilla headquarters on Yongpyong, which was in touch with Yongmae, informed her that there were enemy troops on Kobuksom an island near Yongmae. After forty rounds of 4-inch dispersed these troops, *Athabaskan* had to content herself with illuminating the mudflats for the benefit of the Yongmae defenders. She could not bring her firepower to bear against targets hidden from view, without shore-based spotting and communications.[6]

The defenders were doing well on their own, putting out terrific barrages of mortar, machine-gun, and small-arms fire. A bomber from U.S. Fifth Air Force loaded with flares arrived at 0023, and took over illumination duties from *Athabaskan*. By 0330, the attack had been repulsed, and the enemy troops withdrew to the mainland. The British destroyer HMS *Comus* had arrived in the area at 0230, diverted from her Worthington patrol to support Mu-do. The shelling of it, as suspected, had in fact, been a feint, and no attack developed.[7]

Photo 19-2

Topside area of the British destroyer HMS *Comus* (D43). She is berthed pierside at Kure, Japan, in 1950, "full dressed" in honor of Queen Elizabeth's birthday, on 4 August. Australian War Memorial photograph DUKJ3275

On the night of 11 June, Communist forces again tried to take Yongmae. The light cruiser HMS *Ceylon*, which happened to be in the area, provided the naval gunfire necessary to help the defenders repel the attackers. *Athabaskan*, meanwhile, enjoyed the "fireworks" from her ring-side seat on station off Mudo, and was able to spend a relatively quiet night. Yongmae came under attack for the third time, on the night of 12 June. This time, *Athabaskan* was able to put down some very effective fire, owing to a much more effective communications link between Yongmae, headquarters on Yongpyong, and the ship.[8]

A "flareship" (a plane) arrived, and at once took over illumination duties from *Athabaskan*. It did an excellent job, and with the arrival of a night-fighter and two bombers, a second flareship proved unnecessary. Adequate illumination was being provided, but the night-fighter had no maps of the area, and it proved difficult for them to provide suitable targets for the B-26s. By 0200 (13 June), the enemy was beginning to withdraw, and within half an hour had returned to the mainland. This was their last attempt to take Yongmae during *Athabaskan*'s patrol.[9]

GUERILLA RAID ON THE MAINLAND

That night, 13 June, *Athabaskan* took up a new station northwest of Sosuap-to, deep in a nearby estuary. Her task unit had been reinforced for the night; HMS *Comus* was guarding Yongmae; the ROKN *Bak Du San* (PC-701) was watching Mu-do; and ROKN *Daejeon* (JMS-301) was waiting in reserve near Yongpyong-do. The reason for *Athabaskan*'s position, far forward of her normal operating area, was to enable her to provide gunfire support for a Wolfpack raid on the peninsula north of Sosuap-to.[10]

Some 300 guerillas under a U.S. Marine Corps sergeant were to land near Sulgumdong, push northward across the peninsula, then sweep to the right toward the beaches near Kumsan-ni, where they were to re-embark. *Athabaskan* was to provide fire support during the night, with *Daejeon* joining at first light to render close support. A combat air patrol from the light fleet carrier *Bataan* was also to arrive at daybreak.[11]

The assault force—embarked in twelve sailing junks, towed by motor junks—was to go in at 0300. Owing to challenges, including the vagaries of tide and wind, and the insubordinate temperament of the guerillas, the landing was not made until almost 0600—well after dawn, about a mile from the intended site. While awaiting arrival of the assault force, *Athabaskan* softened up the landing area while Marine Squadron VMA-312 Corsairs circled overhead ready to crush any opposition to the landing. *Daejeon* had also arrived and lay off the beach to lend aid with her 3-inch and 40mm guns, should it be required.[12]

Photo 19-3

Marine Corps F4U-4B Corsair fighter parked aboard USS *Badoeng Strait*'s (CVE-116) snowy flight deck, during operations off the Korean coast, 14 November 1950. Naval History and Heritage Command photograph #NH 97373

After reaching the beach without much difficulty, the guerillas quickly moved inland. An enemy mortar battery which took them under fire, was quickly knocked out by the carrier planes. *Athabaskan* was in radio contact with the U.S. Marine leading the assault (whose code-name was Blackjack). Accordingly, it was possible to follow the progress of the attackers fairly well, except for two short interruptions when Blackjack's ear-phones were knocked off by a sniper, and when a rifle bullet damaged the controls of his radio.[13]

At about 0730, *Athabaskan* noticed that the junks, which had proceeded north toward Kumsan-ni where they were to reembark the guerillas at completion of the raid, were contrary to plan, beginning to withdraw from the beach area. Guerillas were also observed running along the shore toward the junks. *Athabaskan* then in touch with Blackjack, who was preparing to lead an assault on an enemy hill-top position, told the ship that no retreat had been ordered, and that he intended to continue as planned; then his radio went dead.[14]

Apparently, Blackjack was mindful of the information that the junks were preparing to withdraw, because he hurried to the extraction area to take charge. When he got there, it was too late to do anything but order a general withdrawal. Apparently, the invasion craft were all civilian-manned junks pressed into service to support the raid. While waiting for the guerillas to return, some of the skippers had decided that an enemy beach in broad daylight was no place for them to be. There had been no guards purposely left behind to police them. Fortunately, some of the guerillas had not followed their companions inland and spotted the intended actions by the junks in time, and were able to stop them and hold them pending the return of the assault force.[15]

The recalled guerilla force was reboarded. As the sail junks were being taken in tow by motor junks, an enemy 120mm battery began to shell them. Neither *Athabaskan*, nor the Corsairs were able to identify the location of the battery, but fortunately it scored no hits, and the raiding party got away safely.[16]

Despite many setbacks, the guerillas considered the raid to have been worthwhile. They estimated that *Athabaskan* and the carrier planes had inflicted 60 casualties on the enemy, and that they themselves had accounted for another seven killed. The raiders had "rescued" 20 friendly civilians and "liberated" two junks, 35 bags of rice, and other supplies, at a cost to them of only one man slightly wounded.[17]

Marine Squadron VMA-312 aboard *Bataan* flew forty-five sorties that day. Armed reconnaissance Corsair aircraft attacked 200 troops, with unassessed damage. Other losses to the enemy, which may or may not have been associated with the raid, included the destruction of one observation post, and damage to seven gun-positions.[18]

DEFENSE OF CHANGNIN-DO ISLAND

> *For limited offensives up to a few thousand meters [inland], the [guerillas] were very good offensive fighters, because they all knew how to use the bayonet, rifle and hand grenade. Therefore, we were able to carry out some creditable military operations.*
>
> —Lt. Col. Jay D. Vanderpool, commander of the U.S. Eighth Army's guerilla unit in Korea.[19]

At 0944 on 15 July 1952, the light fleet carrier USS *Bataan* received a report from Marine Col. James T. Wilbur, commander, West Coast Island Defense Element (CTE 95.15), of an amphibious invasion of Changnin-do by North Korean forces. This report was accompanied by a request for strike and target combat air patrol support. Her embarked Squadron VMA-312 aircraft assigned to other missions that day were immediately diverted, and the remainder of the day's schedule was revised to provide all possible assistance.[20]

Changnin-do was one of many islands in the Yalu River Estuary, part of a long string of islands originating off northwest Korea that extended down around the peninsula to Pusan in the southeast. Some of these islands were of great tactical importance to Allied forces. Friendly, guerilla-held islands were ideally suited for radar units and signal intercept stations, and also served as bases for other elements. They offered safe haven for helicopter teams and boat crews dedicated to rescuing downed airmen. Islands behind enemy lines, in particular, served as springboards for guerilla actions and agent insertions.[21]

Additionally, guerillas occupied key terrain that controlled several Yellow Sea choke points. Their control of these friendly islands limited enemy movements around the mouth of the Yalu River, into the port cities of Chinnampo and Haeju, and within the important Han River Estuary. Allied control of the sea, gained with guerilla assistance, forced all supply support for the enemy front lines to move overland or by rail, making them vulnerable to air attacks.[22]

That morning (15 July), 156 North Korean Army (NKA) soldiers in two sail junks and four wooden folding boats equipped with outboard motors invaded Changnin-do. The light cruiser HMS *Belfast*, frigate HMS *Amethyst*, and aircraft from *Bataan* assisted friendlies by engaging targets of opportunity on or near the island. In the action, *Amethyst* was taken under fire by a 75/76mm battery but was not hit. The British warships, after engaging in counterbattery fire, which caused secondary explosions, and silenced enemy guns, positioned themselves near the island for future support awaiting a counterattack by friendly forces. (*Amethyst* later took part in the 1957 film *Yangtse Incident: The Story of HMS Amethyst*, depicting the heroics of her crew on the Yangtse River during the Chinese Civil War in the summer of 1949.)[23]

The following day, 16 July, *Bataan* launched forty-four VMA-312 aircraft with their main effort directed toward the retaking of Changnin-do. A Corsair piloted by Marine Capt. Charles L. Duncan, was damaged by gunfire at Changnin-do and forced to ditch off the island. A small boat from *Amethyst* later retrieved Duncan from his life raft, and he was returned by the "Worthington Patrol destroyer" to *Bataan* two days later. (This term refers to each night, one of the screening ships for the carrier being detached to carry out the patrol, and then returning to the carrier group the following morning.) [24]

Friendly guerillas recaptured the island on the 17th, supported by the *Belfast*, *Amethyst*, and South Korean patrol craft *Kum Kang San*. Of the NKA invasion force that had occupied the island, 60 enemy soldiers were killed, 30 drowned while trying to escape, 41 were taken prisoner, and five went missing. Friendly losses were eight killed, and 12 wounded.[25]

Photo 19-4

Kum Kang San (formerly USS *PC-799*) off the Mare Island Naval Shipyard In California, 17 June 1950, following transfer to the South Korean Navy. Naval History and Heritage Command photograph #NH 85482

OPERATION SICIRO

Our mission was to harass and interdict the rear areas. We conducted raids and ambushes and laid mines along the MSRs [Main Supply Routes].

—Maj. Richard M. Ripley, U.S. Army Special Forces, who commanded Wolfpack, a partisan guerilla organization, in the spring of 1952.[26]

Photo 19-5

Water color by Frank Norton depicting motor and sail junks of Wolfpack irregular forces alongside the Australian destroyer HMAS *Bataan* (D191), circa 1952. Australian War Memorial photograph ART40045

During the first two weeks in September 1952, the destroyer HMCS *Iroquois* served as commander, Task Unit 95.12.4 in the Haeju area. The patrol was characterized by considerable enemy activity on the mainland opposite Mudo and by some heavy shelling of the friendly guerillas on that island. However, the principal event was a fairly large-scale raid carried out by Wolfpack guerillas, supported by HMCS *Iroquois* and HMS *Belfast* and planes of the escort carrier USS *Sicily*.[27]

Established in January 1952, WOLFPACK was composed of eight units (initially totaling 3,800 partisans) designated Wolfpacks 1 through 8. Wolfpack headquarters was on the large island of Yongyuong-do, due west of Seoul, with the units on adjacent islands south of the 38th parallel. Wolfpack conducted operations behind enemy lines, northwest of Seoul in the southern portion of the Ongjin Peninsula.[28]

The day of the raid, 10 September, the landing force was in its junks by 0830, to return to Yongmae-do, from which the raid was launched.[29]

The raid, code named Operation SICIRO, involved sending three guerilla companies in motor and sail junks from Yongmae-do (an islet in Haeju Bay) across the shallows to the edge of the mud-flats opposite the Changdong Peninsula. Under the cover of *Iroquois*' 4-inch, twin mount, whose fire was to be directed by a team ashore, the raiders intended to make a quick foray inland at about 0400, and retire to their junks about 0800 with what prisoners they were able to capture. At daybreak, aircraft from *Sicily* were to help cover the withdrawal.[30]

Photo 19-6

Painting by Frank Norton of Wolfpack Headquarters on Yongpyong-do, with HMAS *Bataan*, and U.S. Navy and ROK ships, August 1952. Australian War Memorial photograph ART40011

Iroquois was involved in several raids, in 1952. Some of the photos in this chapter are those of Lt. George MacFarlane, an officer aboard the destroyer. His son John MacFarlane graciously allowed their use, and shared recollections of his father's descriptions of the raids:

The raiders would arrive by junks supported by the destroyer. Apparently, the important part was the covering fire in the extraction. The raiders would put on markings (like same colored T-shirts for example) so that they could be distinguished from Communist forces. On at least one raid the enemy forces also put on the same markings (I guess an agent had discovered the key) and *Iroquois* did not know which forces to attack. My dad said these operations were somewhat chaotic – as plans often either changed during operations or communications were muddled. He said that on at least one operation enemy junks arrived unexpectedly.

Comdr. William Landymore, the ship's captain, was under some pressure to create raiding opportunities. His officers, not trained in these type operations, were understandably not keen about them. They were more comfortable with blockade duties, shooting up shore targets, etc. The USMC and ROK liaison officers were memorable real life "Rambo-type" characters armed to the teeth and fearless.[31]

Photo 19-7

Wolfpack guerillas in a boat alongside HMCS *Iroquois*, and Lieutenant Park, ROKN liaison officer aboard *Iroquois*, circa 1952.
Courtesy of John MacFarlane

Task Element 95.11, comprising escort carrier USS *Sicily* under the command of Capt. Almon E. Loomis, USN, with Marine Attack Squadron VMA-312 embarked, and screened by HMS *Cossack* (D57),

HMCS *Nootka* (DDE213), and USS *Marsh* (DE-699), was operating in the vicinity of 37°30'N, 124°30'E. Its mission included defending the islands of Sok-to, Cho-do, Paengyong-do, Taechong-do, Yongpyong-do, and Tokchok-do.[32]

On the evening of 9 September, the light cruiser HMS *Belfast* (CTE 95.12) arrived off Yongpyong-do on a routine visit, and offered to help support SICIRO. Her 6-inch guns would prove of great value during the operation. *Belfast* and *Iroquois* opened fire at 0230 on 10 September, and for ninety minutes their guns poured high-explosive rounds on enemy positions in the assault area.[33]

At 0400 the guerillas left their junks at the edge of the mudflats, and moved inland. Throughout the assault and during the withdrawal, frequent calls for fire support by *Belfast* and *Iroquois* were answered promptly and effectively. A larger enemy force, pressing hard the withdrawal of a guerilla company, was stopped in its tracks, when two 6-inch shells from *Belfast* dropped in the midst of the attackers.[34]

Photo 19-8

Lt. Earl McConechy, RCN, and Ordnance Lieutenant Percy Buzza, RCN, standing in front of HMCS *Iroquois*' 4-inch guns.
Courtesy of the John MacFarlane

Photo 19-9

Iroquois's commanding officer, Comdr. William Landymore (viewer's left), and principal armament control officer, Lt. George MacFarlane (wearing battle helmet), search for and select gunnery targets during action on the Korean coast.
Courtesy of John MacFarlane

Sicily's aircraft arrived at 0620, and commenced bombing and strafing the disorganized enemy forces. The Wolfpacks were delighted when an aerial bomb scored a perfect hit on an enemy bunker, producing many North Korean casualties.[35]

Photo 19-10

Water color by Frank Norton of aircraft bombing enemy troops in support of a Wolfpack raid, 25 September 1952.
Australian War Memorial photograph ART40052

On 10 September, the day of the raid, Sicily flew 33 sorties, consisting of 12 CAP (Combat Air Patrol), 10 TARCAP (Target CAP), and 13 strike and reconnaissance, in support of her many missions. That afternoon, four MiG-15s attacked a two-plane section of TARCAP in the general vicinity of Cho-do. In a dogfight that followed, one MiG was shot down in flames, and one F4U Corsair was damaged. *Cossack* had been detached at 1700 the previous night for the nightly island patrol. She returned earlier that day at 1000, and at 1750, *Nootka* was detached in rotation to carry out the next patrol.[36]

Photo 19-11

Starboard side view of USS *Sicily* (CVE-118) with F4U Corsairs parked aft, April 1954. Naval History and Heritage Command Catalog #NH 97317

Operation SICIRO was a great success, with the cost to the guerillas of only four men lightly wounded. Reports from friendly agents and prisoners of war estimated about 400 total enemy casualties in dead and wounded. Moreover, several gun positions were neutralized or destroyed, and the enemy defenses in the area thrown into disorder. Three friendly agents captured earlier by the enemy, were rescued and brought back to Yongmae-do.[37]

Perhaps the most important result of the raid was the effect it had on Wolfpack morale. In the four nights after SICIRO, Wolfpack landed more agents in enemy territory than it had in the entire month of August preceding it.³⁸

Iroquois remained in the Haeju area for four days following the operation. She was relieved as commander, Task Unit 95.12.4, by the frigate HMS *St. Brides Bay*, on the morning of 14 September, and sailed to join the screen of the light carrier HMS *Ocean*. *Sicily* had been relieved of her duties by *Ocean* at 2100 on 13 September, after which she left with *Cossack* as escort for Sasebo.³⁹

Photo 19-12

Bow view of HMS *Ocean* at anchor in World War II, location and date unknown. Australian War Memorial photograph 302467

20

HMCS *Nootka* Captures North Korean Minelayer

About 0200 hours I picked up a bogey on the radar and reported it immediately. "Action Stations" was sounded and the Operations Room crew closed up. As this was taking place, I reported that the bogey had split in two and one part disappeared from the screen. McIlvoy MacDonald, the Senior R.P. rating [radar operator] at the time wasn't sure what happened and cautioned me to keep a close watch. Shortly thereafter Capt. Steele entered the Operations Room just as the bogey split once more and disappeared. I reported the same directly to the Captain and McIlvoy.

[Ship's captain] Cdr. Steele asked me what I thought was happening, it came to mind that the disappearing bogeys had to be mines. At first, he seemed perplexed by this, then I mentioned that our ship was about 1000 yards from where the first object disappeared. He gave orders to alter course 20 degrees to Port. Had he not done so, we would have sailed right over where the first object disappeared. I was then relieved of my position which was taken over by the killick of my watch [leading rating], Scotty Morrison.

—Former HMCS *Nootka* crewmember, Donald M. Jatiouk, describing gaining radar contact on an unknown contact, which proved to be a North Korean vessel engaged in minelaying.[1]

Photo 20-1

Ship's Badge

Motto: Ready to Fight

Royal Canadian Navy destroyer HMCS *Nootka* (DDE-213) under way. Image from ship's 1957 Christmas card (http://jproc.ca/nootka/memorabilia.html)

On the evening of 25 September 1952, the Canadian destroyer HMCS *Nootka* (DDE-213) was on her normal station southeast of Cho-do, when she gained radar contact on an unknown craft moving north off the coast of Songhang-ni. Turing seaward, the vessel proceeded toward Cho-do, then stopped in the Cigarette channel between the mainland and the island. *Nootka* had been informed that a friendly junk would be in the area that night, so the unknown craft's presence and movements did not initially cause any concern. But then, it suddenly reversed course, and headed for enemy territory.[2]

Nootka quickly tried to intercept, but the craft had too big a lead, and was able to gain the safety of the shallows near shore. The destroyer's commanding officer, Comdr. Richard M. Steele, RCN, wishing to capture the vessel, and not destroy it, did not order it taken under fire while fleeing. However, an attempt was made to drive the craft out to sea, by firing high explosive rounds into the cliffs against which it had taken shelter, but to no avail.[3]

Thinking that the stranger might have been laying mines out in the Cigarette channel, Steele requested that a minesweeper check sweep the area. The steel-hulled USS *Defense* (AM-317) arrived the following evening, 26 September, and carried out a thorough check. Finding nothing, she left the area at midnight. However, as she was departing, both she and *Nootka* picked up a radar contact near where the junk had sought refuge the previous night. Steele, determined not to let the craft escape a second time, quietly moved *Nootka* into position to intercept if it should again move out into the Cigarette channel.[4]

The junk acted as anticipated, while *Nootka*, desiring to trap it as far out to sea as possible, waited patiently for this expected eventuality. The destroyer gained radar contact at 0200 on an object thought to be the suspected vessel out in the channel, as described by Able-Seaman Radar Plotter Donald M. Jatiouk, in the quoted material at the chapter's head. Jatiouk's account of the incident, continued below, indicates that he was unaware of the commanding officer's belief that the contact was an enemy vessel:

> From there on, the Captain became convinced it was an enemy vessel and took the necessary action to capture it. After starshell and a few rounds of 4 inch to convince the crew to abandon their effort, all of them left the junk in inner tubes. Our seaboat was launched with an armed crew to pick them up. Six were pulled out of the water, one resisted and was shot, and another floated to a barren rock reef. He was picked up in the morning in a state of hypothermia. The abandoned junk was brought alongside in the morning and searched for documents, while we were careful about

booby traps. It was then cast adrift and used for target practice by the different [gun] crews, then sunk.

The next day an American minesweeper, with magnetic sweeping gear, blew up 3 mines to once more make our patrol zone safe. I watched one of the explosions - it was huge. If we had run over one of the mines, I'm certain the ship would have been sunk with lives lost.[5]

What *Nootka* had rapidly closed with turned out to be a large junk with unusually low freeboard. Sighting small dark objects dropping from the vessel's stern, and floating toward her, *Nootka* engineers went astern and created a wash that kept them clear of the ship. The destroyer then drew off to some 1,800 yards distant, illuminated the junk, and fired several rounds of Bofors into it at the waterline—to encourage anyone who might still be aboard to abandon.[6]

An armed party was set off in a motor cutter to investigate the floating objects. As the party approached the nearest one in the gloom, its members saw that it contained what appeared to be a human body with its feet sticking up. Illumination from a flashlight revealed a North Korean officer sitting in a large, inflated truck tire tube and about to open fire with a submachine-gun. Apparently blinded by the light, he missed. The sailors returning fire riddled him and his unconventional floating device, with bullets, and both promptly sank. The remaining enemy crewmen were left floating in their tubes until daybreak, while the junk was towed further out to sea for later examination.[7]

JUNK CHARACTERISTICS / PREVIOUS ACTIVITIES

At dawn, *Nootka* returned to pick up survivors and managed to retrieve all five, two lieutenants and three petty officers of the North Korean Navy. Even though they had all discarded their weapons, the officers did not allow themselves to be taken without a struggle. Warm baths, clean clothing, hot rum toddies, and cigarettes resulted in a marked change in the attitude of the prisoners. In particular, one of the petty officers became very cooperative, chatting with the Canadians, through an ROK interpreter on board, and divulging much valuable information.[8]

The large junk had been commandeered by the North Korean Navy and converted into a minelayer. Changes included the removal of its superstructure, in order that, when fully loaded, a mere 18 inches of freeboard remained, lessening the chance of visual detection. Another stealth measure employed was the use of oarsmen to propel the craft, eliminating the necessity of mounting an engine whose sound might disclose its location.⁹

The cut-down junk was also heavily reinforced to enable it to carry two magnetic mines, weighing approximately one ton each, which were launched by rolling over the stern on wooden rails. The minelayer had only carried out two operational missions (on 19 and 20 September) before being detected the first time by *Nootka*. Thus, the Communists were only able to lay eight mines (two each on 19, 20, 25, and 27 September), which were destroyed by "sweepers" before they could do any damage. The short operational life of the "Cho-do minelayer" must have been a disappointment to the North Korean Navy, which apparently did not repeat the experiment.¹⁰

SOVIET MINES USED IN THE KOREAN WAR

Photo 20-2

Dock landing ship USS *Colonial* (LSD-18) under way in the early 1950s.
Naval History and Heritage Command photograph #NH 107629

During the Korean War, North Korea, with Soviet naval help, managed to emplace thousands of mines to protect the ports of Wonsan, Hungnam, and Chinnampo, and locations of less importance, including waters around northwest islands. These mines were of five types, including a new one discovered at Wonsan. On 11 March 1952, Mine Squadron 3's Mine Disposal Team, assisted by Underwater Demolition Team 5, both embarked in the dock landing ship USS *Colonial* (LSD-18), recovered the new type of Russian mine. Designated MYaM, it was a moored, contact type, laid by surface craft and designed for use in shallow water.¹¹

The Soviets provided the North Korean Navy with two varieties of moored contact mines (M-26, and the new MYaM), two magnetic influence mines (KMD-500, KMD-1000), and one that could be adapted as either a contact or influence mine (MKB). Based on the descriptions provided by survivors of the enemy minelayer, they had been sowing coastal waters with KMD-500 bottom-influence mines.

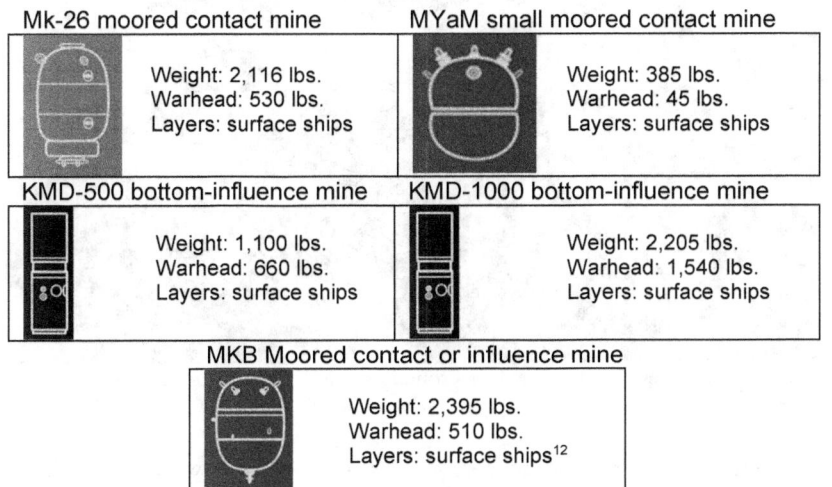

Soviet Mines Supplied to the North Korean Navy

Mk-26 moored contact mine	MYaM small moored contact mine
Weight: 2,116 lbs. Warhead: 530 lbs. Layers: surface ships	Weight: 385 lbs. Warhead: 45 lbs. Layers: surface ships
KMD-500 bottom-influence mine	KMD-1000 bottom-influence mine
Weight: 1,100 lbs. Warhead: 660 lbs. Layers: surface ships	Weight: 2,205 lbs. Warhead: 1,540 lbs. Layers: surface ships

MKB Moored contact or influence mine

Weight: 2,395 lbs.
Warhead: 510 lbs.
Layers: surface ships[12]

NOOTKA'S PREVIOUS "VERY ACTIVE" PATROL

HMCS *Nootka* was very fortunate, in not suffering any personnel casualties during the above-mentioned activity and, for that matter, over the entire course of the war. During an earlier inshore patrol from 19 July to 6 August (while serving as CTU 95.12.4), she came under enemy fire on seven occasions, but each time escaped harm and gave back more than she received. During that patrol period, enemy artillery in the Haeju area was most aggressive and persistent in hammering the friendly islands and the UN naval ships at every opportunity.[13]

Repeatedly, *Nootka* was called upon to silence enemy batteries firing on friendly islands, and often came under fire herself. Planes of TE 95.11 provided splendid support when the weather was suitable, and bombers of the U.S. Fifth Air Force responded quickly to requests for assistance.[14]

Photo 20-3

HMCS *Nootka* crewmembers loading a round into one of the destroyer's 4-inch guns. Library and Archives Canada/Department of National Defence fonds/a213204

The enemy raiding forces, on the other hand, were surprisingly inactive. Not a single amphibious attack was made on a friendly island during *Nootka*'s tour, while UN guerilla forces sent intelligence parties ashore practically every night, and also launched occasional large-scale raids against the mainland.[15]

BAN OF GROG ABOARD COMMONWEALTH SHIPS

The rum ration ("grog") was a long-standing tradition beloved by sailors of the Royal Canadian Navy (and the North Korean prisoners aboard HMCS *Nootka*, in September 1952), but one that no longer fit a modern navy. The Royal Navy led with "Black Tot Day" (31 July 1970), and the RCN issued its last rum ration on 31 March 1972.

Leading up to elimination of the daily tot in the RN, the Admiralty Board had, on 17 December 1969, issued a written answer to a question from a Member of Parliament, Christopher Mayhew, stating:

> The Admiralty Board concludes that the rum issue is no longer compatible with the high standards of efficiency required now that the individual's tasks in ships are concerned with complex, and

often delicate, machinery and systems on the correct functioning of which people's lives may depend.[16]

This led to a debate in the House of Commons on the evening of 28 January 1970, the "Great Rum Debate," which lasted an hour and 15 minutes. The debate closed with a decision that the rum ration was no longer appropriate.[17]

The Royal Australian Navy had discontinued its ration in 1922. Rather belatedly, the Royal New Zealand Navy finally gave up its rum ration, on 27 February 1990.[18]

Photo 20-4

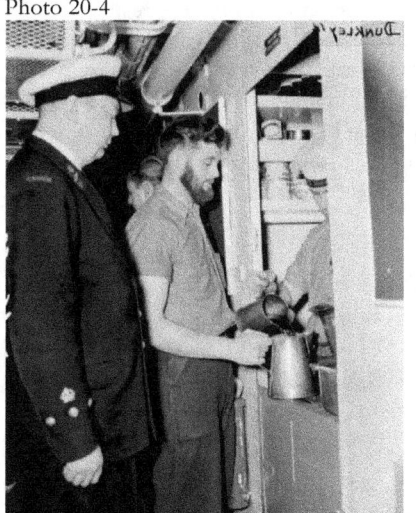 

Chief Petty Officer Joseph Leary, Coxswain of HMCS *Nootka* (left) supervising the preparation of the men's daily grog ration, made from watered-down rum. At the right is a newspaper article, proclaiming the end of the tot aboard HM Canadian naval ships. Australian War Memorial photograph DUKJ4083

HISTORY OF THE TOT

A mightily bowl on the deck he drew and filled it to the brink;
Such drank the Berwick's gallant crew and such the Gods shall drink;
The sacred robe which Vernon wore was drenched within the same;
And hence its virtues guard our shore and Grog derives its name.

—Unknown crewmember of HMS *Berwick*, a ship in Adm. Edward Vernon's squadron. Known as "Old Grog," the admiral derived his nickname from a boat clock made of a coarse material called grogram, which he wore while pacing the deck of his flagship.[19]

"UP SPIRITS" was the age-old call for sailors of the Royal Navy, and later, those of other Commonwealth navies as well, to muster for their daily issue of rum. The well-loved Grog was called thus in reference to its creation in the 1700s by Adm. Edward Vernon, RN (known as "Old Grog"). Vernon, appalled by debilitating effects on "all hands" of the then daily issue of 1/2 pint per day of neat 80 proof rum, plus a gallon of beer if they desired it, issued the following guidance to the commanding officers of the ships in his West Indies Squadron:

> You are therefore hereby required and directed, to take particular care that rum be no more served in soecie [in respect to kind] to any of the ship's company under your command, but that the respective daily allowance of half a pint a man for all your officers and ship's company, be every day mixed with the proportion of a quart of water to every half pint of rum, to be mixed in a scuttled butt kept for that purpose, and to be done upon deck, and in the presence of the Lieutenant of the watch, who is to take particular care to see that the men are not defrauded in having their full allowance of rum.[20]

Vernon's earlier efforts to allay the effects of neat rum by diluting it with lime juice led to British ships becoming known as "Lime Juicers" and British Sailors themselves as "Limeys." As the years went by, further cuts were made in the daily rum issue and by 1850, the midday ration had been reduced to one 2 ½-ounce tot.[21]

The daily rum tot as well as being a social aspect of life on the lower deck, became a medium of barter, a way of paying for a favor, such as standing a watch or doing a shipmate's laundry, or paying off a bet. There were standard currencies for such occasions. "Sippers" was the least, a small sip from a mate's tot. "Gulpers" was one, but only one, big swallow. "Sandy Bottoms" referred to draining off what was left of a tot offered by a shipmate. While not an exact measure:

- 3 Sippers = 1 Gulper
- 3 Gulpers = 1 Tot[22]

The term "wet" had various meanings. These included, almost just the touching of a mate's rum (for a very small favor), or as a general reference to partaking of alcoholic beverages, such as "Higgins went ashore and had several wets."[23]

21

War Drags On

Photo 21-1

Drawing by Hugh Cabot of the tent at Panmunjom (a village near Kaesong) in which armistice talks dragged on for months, 1951.
Naval History and Heritage Command photograph #88-187-DQ

In the ten months between late September 1952 (when HMCS *Nootka* captured the North Korean minelayer), and the end of the Korean War, on 27 July 1953, little changed regarding island defense off the west coast of Korea. There were several incidents of enemy shore artillery fire at islands and ships, with little damage and few casualties resulting. UN ship support of much reduced guerilla activities ashore continued, concurrent with difficulties, and interruptions in the armistice talks.

OCTOBER – DECEMBER 1952

On 8 October 1952, the United Nations senior delegate to the armistice talks at Panmunjom declared an indefinite recess of the senior delegates' meetings. This action followed sixteen months of incessant Communist propaganda tirades, charges, and countercharges in the to-date, futile talks of peace. During this recess, gridlock in the war continued as before, including in the western islands, where the enemy mounted only minor attacks, mostly in the Haeju area, against Cho-do and Sok-to. Those against Cho-do consisted of small aircraft dropping bombs, and shore batteries taking ships guarding the island under fire.[1]

ATTACKS AGAINST CHO-DO ISLAND

Two single-engine slow enemy aircraft made bombing runs on Cho-do, on 13 October, dropping 15 bombs in pairs and widely dispersed. The main point of impact was within 200 yards of the radar station, but there were no casualties or damage to the installation.[2]

Task Unit 95.12.1's main function was defense of Sok-to and Cho-do, the latter island being the site of a radar station and Tactical Air Defense Center, a vital point in the operations in South Korea. In autumn 1952, the task unit consisted of British, Australian, Canadian, and American units: destroyers HMS *Comus*, HMAS *Anzac*, and HMCS *Crusader*, the rocket-armed medium landing ship USS *LSMR-412*, and some small vessels of the South Korean Navy.[3]

Photo 21-2

Medium landing ship, rocket USS *LSMR-412* under way.
National Archives photograph #USN 1045120

Individual replacement ships periodically joined or departed task units, as they began or completed patrols. On 3 November, the minesweeper USS *Condor* (AMS-5) was taken under fire by two shore batteries of two guns each in the Cho-do area. There were no casualties or damage resulting from 50 rounds fired by the 75mm-105mm guns. Two weeks later, on 16 November, the destroyers USS *Lyman K. Swenson* and HMAS *Anzac* received an unreported number of rounds from four 75/76 mm guns in the vicinity of Cho-do. There were no casualties or damage suffered.[4]

On 26 November, six single-engine unidentified aircraft dropped five bombs on Cho-do, with no damage reported. Ten days later, on 6 December, four enemy aircraft dropped two bombs on Sok-to and ten on Cho-do. There were again, no casualties or damage reported. HMCS *Crusader* fired 79 rounds at the aircraft with no hits observed. It should be noted that anti-aircraft guns in that era were designed for use in defense against aircraft attacking the host ship. Range of the aircraft from the destroyer would determine whether or not hits were likely.[5]

Photo 21-3

HMCS *Crusader* (DDE228) under way off Korea, 3 March 1954.
National Archives photograph #80-G-642747

On 23 December, Cho-do received approximately 125 rounds of enemy fire (caliber unreported), with no resulting personnel casualties or material damage to the island. Two hundred rounds also struck the eastern side of Sok-to, but also caused no harm there.⁶

HAEJU-ONGJIN AREA

Other incidents, in Autumn 1952, took place further south in the Haeju-Ongjin area. On 19 October, shore batteries in the western Haeju approaches fired on friendly junks, following which HMS *Morecambe Bay* silenced the enemy guns with 24 rounds of counterbattery fire. Two 75/76mm guns in the Haeju area took the ROKN patrol craft sweeper *Hwa Seong* under fire on the 31st. Some of the 35 rounds landed within 10 yards of the ship, but she was not damaged. Other rounds fell on Losuap-to Island. HMCS *Nootka* silenced these guns.⁷

On 15 December, the friendly guerilla-held island of Mudo in the Haeju approaches, received 30 rounds of 75/76 mm fire, but suffered no casualties. Eight days later, as the ROKN motor torpedo boats *Galmaegi* (PT-23) and *Gireogo* (PT-25) fired rockets at a village on the western Ongjin peninsula, they were taken under fire by machine guns and a mortar. There was not damage to the PT boats.⁸

Photo 21-4

Personnel of ROK Navy stand at attention on board a PT boat while Korean national anthem is played. PT-23, PT-25, PT-26, and PT-27 were added to ROK Navy when Vice Adm. Sohn Won Il, ROKN, accepted the vessels at Sasebo, Japan.
National Archives photograph 80-G-438008

JANUARY – APRIL 1953

During the first part of 1953, the enemy continued to harass friendly-held areas and Allied defense forces in the disputed areas. Cho-do received the brunt of the assault, along with Haeju, much as they had in the frigid last months of 1952. UN forces continued to defend these islands, incurring casualties.[9]

CHO-DO ISLAND

In early 1953, as in late 1952, Cho-do was the enemy's favorite target for shore bombardment from the mainland. Almost all of the gunfire was inconsequential, except for an attack on 4 April, in which there were eleven casualties. Information in the table includes the size of enemy guns, number of rounds fired, and damage/casualties inflicted.

Enemy Shore Artillery Attacks on Cho-do

Date	Guns	Rounds	Results
15 Jan	75/76mm	7	No personnel casualties or material damage
14 Feb	unreported	2	No personnel casualties or material damage
2 Apr	75/76mm	4	No personnel casualties or material damage
4 Apr	75/76mm	220	One man killed and 10 wounded
13 Apr	unreported	9	No personnel casualties or material damage[10]

HAEJU AREA

The first month of 1953 in the Haeju area brought enemy artillery fire both at friendly islands and on USN and ROKN minesweepers. On 15 January, Mudo in the Haeju approaches received shore fire, with no damage or casualties reported. That same day, while carrying out minesweeping operations in the Haeju area, USS *Pelican* received 10 rounds of estimated 75/76mm enemy shore fire. She was unscathed.[11]

This pattern repeated itself in the latter half of January. On the 18th, the ROKN minesweeper *Kil Chu* (YMS-514) received nine rounds of estimated 75/76mm enemy fire from a shore battery in the vicinity of the Haeju approaches, with no harm accruing. Mudo came under much heavier fire, on 30 January, 120 rounds of estimated 75/76mm enemy fire from batteries on the mainland. No damage or casualties resulted.[12]

ROKN MOTOR TORPEDO BOAT OFFENSIVE ACTION

While UN forces continued to defend islands, ROKN naval units periodically carried out small-scale operations against enemy positions on the mainland. On 8 February, the motor torpedo boats *Galmaegi* (PT-23) and *Gireogo* (PT-25) carried out 5 inch-rocket attacks against enemy positions northeast of Paengyong-do. Return machine gun fire failed to score any hits on the MTBs.[13]

DUTCH FRIGATE CREWMEMBER KILLED

On 26 February 1953, a boat from the Dutch frigate HNMS *Johan Maurits Van Nassau*, while en route to a rendezvous with a junk from Sosuap-to (an islet in Haeju Bay) to pick up a sick South Korean serviceman, was fired on from nearby Taesuap-to. Telegraph operator C. van Vliet was killed, and a ROKN liaison officer seriously wounded, by friendly fire. He was the only member of the crews of the four Dutch warships that served in the war—destroyers HNMS *Evertsen*, *Piet Hein*, and *Van Galen*, and frigate *Van Nassau*—to be killed in action.[14]

The frigate had relieved the destroyer *Piet Hein*, on 18 January 1953, as the Royal Netherlands Navy ship in theater.[15]

Photo 21-5

Watercolor by Frank Norton of the Australian frigate HMAS *Condamine*, questioning the crew of a Korean power junk in Haeju Gulf, Korea. Australian War Memorial photograph ART40050

STALIN'S DEATH SPURS RENEWAL OF TALKS

On 5 March 1953, Joseph Stalin, head of the Soviet Government in Moscow, died of heart failure. His death had a positive impact on the Korean armistice negotiations. The Soviets chose Georgy Malenkov as his successor. At Stalin's funeral, the less-hawkish Malenkov provided political maneuvering space to Mao Zedong and Kim Il Sung (the leaders of Communist China and North Korea) by stating international disputes could be "settled peacefully on the basis of mutual agreement between the countries concerned."[16]

The UN received notification, on 28 March, that China and North Korea were ready to resume negotiations at Panmunjom. On 11 April, UNC and Communist negotiators met at Panmunjom and agreed to exchange their sick and wounded POWs. During Operation LITTLE SWITCH, which lasted from 20 April to 3 May, the UN returned 6,670 communist POWs (5,194 North Koreans, 1,030 Chinese, and 446 civilians). The Communists released 684 prisoners (471 South Koreans, 149 Americans, 32 British, 15 Turks, and several members each from Australia, Canada, Colombia, Greece, Netherlands, Philippines, and South Africa). Little Switch set in motion the diplomatic momentum needed to resume armistice discussions.[17]

APRIL – JUNE 1953

In spring 1953, periodic enemy shore artillery and aircraft bombardment of friendly islands continued, as did enemy artillery fire directed at Task Group 95.1 ships operating within range of the shore guns.

CHO-DO / SOK-TO / HAEJU AREA

Photo 21-6

Drawing by Frank Norton of the Cho-do coastline from the deck of HMAS *Warramunga*. The ROKN landing support ship *Yeongheungman* (LSSL-107) is visible in the distance, 1952.
Australian War Memorial photograph ART40021

On 5 April, four enemy guns on the mainland east of Cho-do, opened fire on Sosa-ri with about 300 rounds of 76 mm shells. One South Korean was killed and ten wounded. Three days later, Sok-to was bombarded, on the 8th, by 60 rounds of unestimated caliber fire from the mainland, with no resultant damage or casualties experienced.[18]

On 10 April, the ROKN landing support ship *Yeongheungman* evaded 20 rounds of unknown caliber enemy shore fire. No casualties or damage to her occurred.[19]

Five days later, on 15 April, an unknown number of unidentified enemy aircraft, apparently flying too low for effective interceptors, bombed Cho-do. Resultant casualties were two killed, 22 wounded.[20]

ENEMY FORCES INVADE YONGWL-DO
Earlier in the month, on 4 April, approximately 90 North Korean Army troops invaded Yongwl-do, near the Haeju approaches, supported by gunfire from the southern Ongjin Peninsula. Aircraft from Task Unit 95.1.1 came to the aid of the besieged island and, after ten hours, the enemy withdrew to the mainland.[21]

GUERILLA RAIDS ON THE MAINLAND
On 3 May, HMS *St. Brides Bay* and the ROKN patrol craft sweeper *Su Seong* supported guerillas infiltrating the mainland in the Haeju area. Five days later, *St. Brides Bay* backed a partisan raid ashore at the mouth of Haeju Bay. The raiders failed to reach their objective and withdrew.[22]

SPORATIC ENEMY ISLAND ATTACKS CONTINUE
On 6 May in the vicinity of the Haeju approaches, enemy batteries fired 20 rounds of estimated 76mm at Mudo and the ROKN patrol gunboat *Chungmugong*, but caused no casualties. There was also no damage or personnel casualties resulting from 130 rounds of 75/76 mm fired at the Canadian destroyer HMCS *Haida*, and 15 rounds at the ROKN minesweeper *Kil Chu*, on 8 May in the Haeju area.[23]

Enemy gunnery in the Haeju area was better, on 15 May, when the ROKN minesweeper *Ganggye* sustained slight structural damage and five minor personnel casualties from a 76mm hit.[24]

On 20 May, the west coast islands of Sangchwira-do and Hachwira-do were targets for 25 and 12 rounds of enemy artillery fire, respectively, but escaped unscathed.[25]

CHINNAMPO AREA

Photo 21-7

HMS *Newcastle* at anchor in Plymouth Sound, England, in World War II.
Australian War Memorial photograph 044941

On 27 May 1953, while firing at enemy gun positions near the approaches to Chinnampo, HMS *Newcastle* drew 30 rounds of 76/105mm return fire. Several enemy salvos straddled the British light cruiser, but no hits were recorded. Two days later, *Newcastle* and the frigate HMS *St. Brides Bay*, while operating west of Chinnampo, were harassed by 30 rounds of estimated 105mm enemy fire. Six rounds fell about 20 yards from *Newcastle*, but no damage or casualties were experienced by the ships. On 3 June, the frigate HMS *Morecambe Bay* was the target of 10 rounds of enemy artillery fire near the approaches to Chinnampo, but the ship escaped damage.[26]

On 3 June, Cho-do was attacked during the evening by three Soviet-built Polikarpov PO-2 biplanes. The eight light-fragmentation bombs they dropped caused no damage or casualties.[27]

22

Evacuation of the West Coast Islands - Operation PANDORA

Within ten (10) days after this armistice agreement becomes effective, withdraw all of their military forces, supplies, and equipment from the rear and the coastal islands and waters of Korea of the other side....

The term "coastal islands," as used above, refers to those islands, which, though occupied by one side at the time when this armistice agreement becomes effective, were controlled by the other side on 24 June 1950; provided, however, that all the islands lying to the north and west of the provincial boundary line between HWANGHAE-DO and KYONGGI-DO shall be under the military control of the Supreme Commander of the Korean People's Army and the Commander of the Chinese People's volunteers, except the island groups of PAENGYONGDO..., TAECHONG-DO..., SOCHONG-DO..., YONGPYONG-DO..., and U-DO..., which shall remain under the military control of the Commander-in-Chief, United Nations Command.

All the islands on the west coast of Korea lying south of the above-mentioned boundary line shall remain under the military control of the Commander-in-Chief, United Nations Command.

—Excerpts from the Korean War Armistice Agreement[1]

As a result of an earlier agreement, on 3 February 1952, regarding the territorial water demarcation line, the UN Command had largely withdrawn from friendly-held islands on the western and northern side of the Hwanghae and Kyong-gi Provincial line. The exception was five islands, namely: Paengyong-do, Techong-do, Sochong-do, Yongpyong-do, and U-do. In accordance with the subsequent Korean War Armistice Agreement, the evacuation had to be completed within ten days after its signing, on 27 July 1953.[2]

West coast islands to be evacuated included the previously defended and garrisoned islands of Cho-do and Sok-to, and seventeen minor islands which were held by the guerillas and, in many cases, were also home to inhabitants and refugees who wished to leave before the Communists took over.³

Photo 22-1

Quiet life of west coast fishing village.
Courtesy of John MacFarlane

On 5 June, the UNC (UN Command) announced that the terms of the armistice would prohibit any civilian from crossing to the UN side of the armistice line who had not been a resident south of it prior to 25 June 1950. This directive meant that the evacuation of the guerillas and their dependents, refugees, and all the inhabitants who wished to move, had to be completed before the signing of the ceasefire. Accordingly, ships of Task Group 95.1, carried out an earlier-than-expected naval evacuation, commencing on 11 June.⁴

In preparation for the evacuation by UN ships of personnel from larger, designated islands, departure of guerillas began early. Following the signing of a POW agreement on 8 June, making a possible armistice imminent, Brig. Gen. Archibald W. Stuart, USA (commanding general, Combined Command, Reconnaissance Activities, Korea), directed each Partisan Infantry Regiment (PIR) Headquarters to begin the evacuation, starting from the outer islands. The 6th PIR, located on Cho-do, Sok-to, and islands farther north, began to withdraw on 10 June using their own boats. The next day, to the south, the 1st PIR in the Paengyong-do area, and the 5th PIR in the Haeju area, also started to leave their small islands.⁵

During the first phase of the evacuation, refugees and inhabitants were also moved. They were instructed to gather on the larger islands such as Paengyong-do, Techong-do, and Yongpyong-do. By 16 June, other than the so-called "stay behinds," small patrol units remaining in place to observe enemy activity, all the guerillas had been evacuated from the outer islands. Although this evacuation was conducted by fishing junks under the command of PIR commanders, and tank landing ships, all loading and landing of individual craft was supervised by Task Group 95.1 ships.[6]

William Henry Cook, former senior communication rating aboard HMS *Modeste*, later described the role the British frigate played in the evacuation, and his part in setting up the necessary radio networks:

> Before these [peace] talks could be started, the UNPIK forces [United Nations Partisan Infantry Korea, also known as the White Tigers] had to be moved off the Islands and got south of the bomb-line [armistice line]. This was Operation 'Pandora' and we were in charge of evacuation of some of the islands. Our seamen were in their element, getting adorned in webbing and collecting rifles, 88 Set [VHF manpack wireless radio], Lanchesters [submachine guns] and Brens [light machine guns].
>
> I had to provide one landing-party operator and he went ashore with the complete set of the 62 equipment [Short-range vehicle-mounted HF set], including the bicycle-battery charger, climbing-irons, Lanchester, and his own webbing and Webley pistol. Eventually, for good all-round-communications, we gave the cox'ns of the boats type 88 sets and tied the 31 set [manpack radio] on the top of the mast and this gave the officer-in-charge ashore, all that he needed so my operator, returned his gear and left it to the cox'ns of the boats and the chap in charge ashore.
>
> It was a sad development that, as the Koreans mustered on the beach to join their boats (landing-craft) the nearby Communist guns lined up on them and commenced Brento shell [firing on] the beach. There were no casualties... As all this was happening, CTU 95.1.2 was supposed to take-over our...part of the coast for the final stages. His ship was delayed so, for a week, our skipper...took over the West Coast. The Yeoman and myself did not turn-in properly for at least a week, but somehow, we did the Cruiser's job.[7]

Other RN ships supporting the evacuation in June, included the light cruisers HMS *Newcastle* and *Birmingham*, and the frigate HMS *Crane*.[8]

On 14 June, an enemy gun battery in the Pungsam district shelled the USS *LST-529* off Cho-do, straddling her four times but scoring no hits. The tank landing ship returned fire with her aft 40mm guns until able to open the range. She suffered no damage. That afternoon she evacuated refugees from Cho-do. Two days later on the 16th, all civilian personnel were evacuated from the villages on Yo-do and Yang-do, as well as the supplies and equipment that were considered excess to immediate needs of the patrol forces remaining behind.[9]

In preparation for a possible armistice, a total of 19,425 persons and their belongings were lifted from west coast islands north of the 38th parallel to southern islands. Before leaving, the inhabitants razed their villages and slaughtered their cattle to prevent them benefiting expected enemy occupants of the islands.[10]

SOUTH KOREA RELEASES COMMUNIST POWS IN VIOLATION OF PROPOSED ARMISTICE TERMS, AND ENEMY FORCES INVADE ISLANDS

Photo 22-2

Aboard the battleship USS *Missouri* (BB-63), Vice Adm. Joseph J. Clark, commander, Seventh Fleet, accompanies Republic of Korea President and Mrs. Syngman Rhee as they visit the ship, off Pusan, Korea, 20 November 1952.
National Archives photograph #80-G-641315

Making the hard negotiations to end the war in progress at Panmunjom even more difficult for U.S. lead negotiators was the fact that ROK President Syngman Rhee was unwilling to accept any armistice at all and so forego the last chance of forcible Korean unification by UN forces. To attempt to derail an agreement, his government released over 27,000 anti-repatriation North Korean POWs from prison camps, on 18 June, in an effort to cause the Communists to break off talks in progress. The Chinese and North Koreans were reluctant to accept the fact that many of their nationals (held in Allied prison camps) did not wish to return to the purported Communist paradise, and the Rhee government continued to decline the very idea of armistice.[11]

As continued armistice talks were postponed, the UNC directed the reoccupation of the outer islands to prevent them falling into enemy hands. Steps were taken to redeploy a small number of guerillas to the outer islands, and provide them with boats so that they could make patrols and withdrawals freely.[12]

Photo 22-3

Australian frigate HMAS *Culgoa* (F408) berthed in port.
Australian War Memorial photograph HOBJ0425

On 23 June, Chinese Communist troops from the Haeju area invaded Yongmae-do, held by a small garrison force. Fortunately, the Australian frigate HMAS *Culgoa* was nearby. After firing her last shots of the war, a 37-round bombardment at the enemy forces, she evacuated the friendly forces. Two days later, the Chinese withdrew from the island and the friendly garrison returned.[13]

A month later, approximately 150 North Korean Army troops attacked Ohwa-do, on 21 July. Arriving offshore in eight large junks, the enemy swarmed ashore and succeeded in killing seven officers and wounding 20 men while establishing a position on the island. The New Zealand frigate HMNZS *Hawae* arrived about four hours later. She opened fire on the enemy soldiers, killing and wounding many, and then successfully evacuated the surviving friendlies.[14]

ARMISTICE SIGNED

Photo 22-4

Light blue United Nations official badge worn by all UN service and civilian personnel in the truce signing area at Panmunjom, on 27 July 1953.
Australian War Memorial photograph P04641.058

On 25 July, the commanding officer of HMS *Birmingham*, Capt. Charles W. Greening, RN, received a message indicating that the signing of a truce agreement was imminent, which signaled commencement of the final west coast island evacuation. Two days later, on 27 July 1953, the armistice was signed at Panmunjom at 1000, with hostilities to cease that night at 2200.[15]

The armistice allowed ten days for this action, but it was decided to complete Operation PANDORA in five days, because of follow-on commitments of the four tank landing ships retained to transport the North Koreans off the islands.[16]

Photo 22-5

Friendly-held west coast island.
Courtesy of John MacFarlane

The first priority was the withdrawal of the redeployed guerillas and refugees from the outlying islands. Under the direction of HMS *Birmingham*, the operation proceeded smoothly. The evacuation began, on 26 July, with guerilla elements in the Haeju area, followed by those in the Paengyong-do and Cho-do – Sok-to areas. All were withdrawn, by 27 July, and the evacuation Sok-to and Cho-do themselves, which were to be handed over to the Communists, commenced. By 1 August 1953, the evacuation of the west coast islands was completed.[17]

Following the completion of PANDORA, HMS *Birmingham* and HMCS *Iroquois* remained for a time in the Cho-do area to provide radar surveillance for two control officers from the Tactical Air Defense Center on Cho-do. They were evacuated, on 6 August, and subsequently the light carrier HMS *Ocean* took over as the Officer in Tactical Command. In order to remind the Communists that UN naval forces were still on the alert, during August, air and surface patrols were continued outside the three-mile limit of the agreed Communist coast.

Ocean conducted three more patrols in Korean waters, up until 16 October.[18]

The British Commonwealth Naval forces remained within a close distance of the Korean coastal waters, until mid-November 1953. This force included a light aircraft carrier, cruiser, and destroyers/frigates. There being no recurrence of hostilities, Rear Adm. Eric G. A. Clifford (Flag Officer Second-in-Command, Far East Station) decided to haul down his flag at Hong Kong, on 18 November 1953. With this occurrence, the deployment of the British Commonwealth Naval forces in the Korean War formally came to an end. A few ships remained for peacekeeping purposes; HMCS *Sioux* did not depart the theater for return to Canada, until 7 September 1955.[19]

Photo 22-6

Private Terry Mahoney, 3rd Battalion, The Royal Australian Regiment (3RAR), guards the entrance to the Demilitarized Zone (DMZ) through the two-strand barbwire fence which extends over the 160-mile battle front. A sign beside him says in Chinese and English, "South Limit Demilitarized Zone, Do Not Enter."
Australian War Memorial photograph HOBJ4541

Postscript

Commodore Dacre Henry Deudraeth Smyth, AO RAN

by Commodore Hector Donohue AM RAN (Rtd)

Commodore Dacre Henry Deudraeth Smyth AO was born in London on 5 May 1923, with his family emigrating to Australia in 1925. Dacre joined the RAN in 1940 and saw service at the Battle of the Coral Sea, bombarding Sword Beach during the D-Day landings in France and was off the Japanese coast when the Atomic Bomb was dropped on Hiroshima on 6 August 1945. Smyth was Executive Officer HMAS *Bataan* during the Korean War and commanded HMAS *Supply* during the Vietnam War. He retired with the rank of Commodore in 1978. In retirement, he became well known for landscape and seascape painting, publishing books containing his paintings and poems. He passed away at his Toorak home, aged 85, in 2008. The information on his career came from Naval Records and the University of NSW *Australians at War* Dacre Smyth Podcast no 1348.

Photo Postscript-1

Captain D. H. D. Smyth RAN Commanding Officer HMAS *Supply* 1967. (AWM)

Dacre's father was the British military hero General Sir Nevill Smyth, who won a Victoria Cross under Lord Kitchener at the Battle of Omdurman in Sudan in 1898 and was considered unlucky not to have been awarded a second VC during the Boer War. He went on to command the First Australian Brigade at Gallipoli and in 1916, the Second Australian Division. His links to Australians led him to emigrate with his family in 1925 and settle at the merino sheep farm Kongbool, near Balmoral in Victoria's Western District. As well, Sir Nevill's first cousin, Lord Robert Stephen Smyth Baden-Powell, was a hero of the 217-day siege of Mafeking in the Second Boer War from October 1899, and later founded the Boy Scout movement, in which Sir Nevill was to have a leadership role in Victoria.

SERVICE IN THE SECOND WORLD WAR

Dacre Smyth joined the RAN in September 1940 as a Special Entry cadet midshipman and after six months' training at the Royal Australian Naval College (RANC), was posted as a midshipman to the heavy cruiser HMAS *Australia*, joining in Sydney in May 1941. *Australia* was then attached to the Royal Navy (RN) and operated in the East Indies Station where the ship was mainly involved in convoy escort duties in the Indian Ocean. She returned to Sydney in December 1941 when Smyth was promoted to Sub Lieutenant. *Australia* became Flagship of the Australian Squadron and was Flagship of the Support Group during the Battle of the Coral Sea when, in May 1942, Smyth celebrated his 19th birthday.

Photo Postscript-2

Commodore Dacre Smyth's depiction of HMAS *Australia* under attack during the Battle of the Coral Sea

One of his abiding recollections of the battle, as his ship was avoiding torpedoes and bombs dropped from Japanese aircraft, illustrates some of the black humour typical of Australian servicemen under fire. He was in the cramped transmitting room just below the waterline passing instructions to the 8-inch gun turrets on the ship when he noticed a wooden mallet swinging by the exit ladder.

He asked his superior what the mallet was for, only to be told: 'When we're sunk, the compartment above will get flooded, so nobody will be able to open the hatch and let us out. We will have enough air in here to survive for a couple of hours while we're sitting on the bottom. In that time some of you might go around the bend. That mallet is for me to knock out anybody who cracks up.' The mallet suddenly looked far more sinister to the 19-year-old than what the Japanese were serving up!

In July 1942 he travelled to England to undertake the normal courses in Navigation, Gunnery, Torpedoes and Signals for young officers in the RN and RAN. Following First-Class passes in these courses, and unable to get an immediate passage back to Australia, he served for a short time in Motor Torpedo Boats based at Lowestoft, Suffolk on the North Sea. Now a Lieutenant, he re-joined *Australia* to complete his training. During this period the ship was involved in bombarding enemy-held islands in the South West Pacific, prior to allied assaults. He gained his Bridge Watchkeeping Certificate in October 1943.

In November 1943 he was posted for exchange service in the RN and joined the World War I Class D cruiser HMS *Danae* in UK waters as gunnery control officer in March 1944. He was just 21 when he took part in the D-Day landings at Normandy, France. On 6 June 1944, *Danae* operated off Sword Beach bombarding the German Batteries near the seaside town of Ouistreham, on Sword's eastern extremity. When the ship had run out of ammunition it was designated as a 'mother ship' for small craft in the Sword area.

Smyth's Commanding Officer, Captain John R.S. Haines RN, told him to get a boat and take his gun crews ashore to see if they could help the army. He added that if he got anywhere near a village or a shop, to purchase a camembert cheese. Smyth asked 'What?' and was advised that it was a sort of French cheese made in Normandy. When he returned onboard that night, he told his Captain that he had the cheese. Haines then called a motor torpedo boat alongside, gave the coxswain the camembert and it was despatched to Portsmouth at 50 knots.

His Captain later explained that he had attended Royal Naval College, Osbourne with King George VI (then Prince Albert, Duke of York) and that they were friends. He had called on the King at Buckingham Palace before sailing and was asked to get a camembert cheese whilst off Normandy. It was the King's favourite cheese and he hadn't tasted one for five years. Smyth's Captain told him that the cheese was on its way back to Portsmouth where a Buckingham Palace car would wait to take it to the Palace for the King's breakfast. Some three weeks later Captain Haines was made a Commander of the Order of the British Empire for his work at Normandy. Smyth served in *Danae* until 3 November 1944, and whilst awaiting passage, flew several missions in the back seat of RAAF 455 Squadron Beaufighters as they attacked German shipping along the Norwegian and Dutch coasts.

In late November 1944 Smyth travelled to Ceylon (today Sri Lanka) to join the destroyer HMAS *Norman* which was operating with the British East Indies Fleet and, from April 1945, with the British Pacific Fleet screening RN carriers. *Norman* was some 85 nautical miles east of Japan on August 6, 1945, when another officer called him on deck to see a 'spectacular sunset.' Smyth immediately wrote a poem, of which the second verse says:

> No cloud, I say, but yet the sun did light
> On towering columns all unreal yet huge
> Which waved and shuddered in grotesque delight
> In myriad hues of ghostly subterfuge.

Unknowingly, he was describing the aftermath of the dropping of an atomic bomb on Hiroshima. 'It had a really shimmery effect, which was the dust in the distance from the atom bomb,' he said. 'We had no idea until the next morning; we'd never heard of an atom bomb.'

POST-WAR SERVICE

Dacre Smyth remained in the Navy after the war and, in December 1945, was appointed as Executive officer of the *River*-class frigate HMAS *Murchison* on commissioning. *Murchison* became part of the British Commonwealth Occupation Force in Japan before returning to Australia in mid-1946 and undertaking training duties. In March 1947 he was given his first sea-going command, the corvette HMAS *Latrobe*, which operated out of Flinders Naval Depot, Victoria taking recruits and Cadet-Midshipmen on training cruises. Following his time in *Latrobe*, Smyth was selected as the Aide-de-Camp to the Governor General Sir William McKell in April 1948.

KOREAN WAR DUTY

Photo Postscript-3

Commander William Marks DSC RAN (left), Lieutenant Dacre Smyth RAN (right) and Lieutenant Patrick Burnett RAN (in background on the compass platform) standing on the wings of the bridge of HMAS *Bataan* as the ship slips from alongside a depot repair ship in the port of Kure, Japan, to take part in the U.S.-led landing at Inchon. (AWM)

After a year in the position, Dacre Smyth felt he had better return to sea to be brought back down to earth as he had begun telling Admirals and Ministers of State what to do. In March 1949 he joined HMAS *Bataan* as Executive officer and again served with the British Commonwealth Occupation Force in Japan. In late June 1950, *Bataan* was en route to Japan for a fifth tour when the Korean War started. From early July 1950 until 29 May 1951, the destroyer operated off Korea; patrolling and blockading, escorting aircraft carriers, and bombarding shore targets.

On 4 December 1950, HMAS *Bataan*, commanded by Commander William B. M. Marks, DSC RAN, was ordered to proceed to the Taedong River Estuary and join her sister ship HMAS *Warramunga*, three Canadian destroyers HMC Ships *Cayuga*, *Athabaskan* and *Sioux* and USS *Forrest Royal*. Captain Jeffry V. Brock, DSC RCN in *Cayuga* was in command as they were ordered to assist in the evacuation of troops and wounded, waiting at Chinnampo some forty miles away.

Captain Brock decided upon night sailing, and the ships moved slowly in line ahead, through icy darkness and falling snow, through the minefields of swept channels. He had warned the passage might be 'fairly tricky.' 'Fairly tricky? That's an understatement if ever I heard one,' wrote Dacre Smyth some years later: 'A winding channel a bare cable's width through minefields, rocks and reefs, with the land far beyond accurate radar range, no lights, a fifty-year old chart, and it is only "fairly tricky." Well, well!'

REMAINDER OF CAREER

In February 1951 Smyth was appointed to the Royal Australian Naval College (RANC) to supervise the new scheme of Intermediate Entry Cadet Midshipmen and shortly after, in April, was promoted to Lieutenant-Commander. He remained at RANC until October 1953 when he joined the *River*-class frigate HMAS *Hawkesbury* in command.

Photo Postscript-4

Copy of a sketch by Lieutenant Commander Smyth of the frigate HMAS *Hawkesbury* (F363) in Buka Passage, Solomon Islands. (AWM)

For the next 14 months *Hawkesbury* undertook routine patrols and training exercises off the Australian and New Guinean coasts including policing Japanese Pearling Fleet operations. She also completed two patrols of Australian waters in the South-West Pacific area. *Hawkesbury*'s tour of duty on the northern patrol ended in December 1954, when she returned to Sydney and paid off on 14 February 1955.

After a year as Staff Officer (Operations) to the Flag Officer in Charge Eastern Australia Area, Smyth was promoted to Acting Commander and sent to Navy Office as Director of Tactics and Staff Requirements in June 1956. He was confirmed in the rank of Commander in December that year and remained on the naval staff until July 1958 when he attended the United States Armed Forces Staff Course in Norfolk, Virginia, graduating in January 1959. This was followed by two years exchange service in the RN where he served as Executive officer of the RN Air Station Sanderling near Glasgow, Scotland.

Returning to Australia in early 1961 he was posted to Navy Office as Deputy Director of Manning and Training, becoming the Director in July 1962 as an Acting Captain. He was confirmed as a Captain in December 1963 and was posted as Naval Officer in Charge Jervis Bay and Commanding Officer HMAS Creswell (the Naval College) in January 1964. In November 1965 in the rank of Commodore he returned to UK as the Australian Naval Representative UK, in the Australian High Commission, London.

After two years he returned to Australia and reverted to Captain to command the *Tide*-class Fleet Tanker, HMAS *Supply*, in December 1967. During his period in command *Supply* participated in a broad range of Commonwealth and SEATO exercises during which time she visited numerous South East Asian ports. She also routinely refuelled HMAS *Sydney* and her escorts deploying to or returning from Vietnam. In 1969, in a first for a RAN replenishment ship, *Supply* was awarded the prestigious Duke of Gloucester's Cup being deemed the RAN ship displaying the highest level of overall proficiency in the Fleet.

PROMOTION TO FLAG RANK

In August 1970 Smyth was posted to the Personnel Division in Navy Office and led a team to review the whole of the RAN's sailor structure, reporting in early 1971. In April 1971 he was promoted to Commodore and posted as Naval Officer in Charge, Victoria and Commanding Officer of HMAS Cerberus at Flinders Naval Depot, the RAN's major training establishment. He remained there until January 1974 when he was posted Director General Naval Personal Services in Navy Office before returning as Naval Officer in Charge Victoria in November 1975. He remained in this position until his retirement in May 1978.

RETIREMENT

Painting was a keen hobby during his time in the RAN, and when he retired, he transformed himself into a compelling artist, as a painter and maker of stained-glass windows, and author. He sold more than 2,800 paintings and self-published 14 books illustrated with his paintings and verse.

In 1977, he was made an Officer of the Order of Australia. In 1994, on the 50th anniversary of D-Day, he was made an officer in the French *Order National du Merite* for his service at Sword beach, and in 2004 he was granted the *Legion d' Honneur* from former French president Jacques Chirac at the 60th anniversary of the D-Day landings.

Photo Postscript-5

Painting of Commodore Dacre Smyth, AO RAN.
Reproduced with kind permission of the artist Ms Irene Hill

Appendix A: ROKN Ships Acquired Before and During the Korean War (78 total)

Tacoma-class Patrol Frigates – 5

Ship	Formerly	Comm.
Dumon (PF-61)	ex-USS *Muskogee* (PF-49)	5 Nov 50
Apnokkang (PF-62)	ex-USS *Rockford* (PF-48)	5 Nov 50
Taedong (PF-63)	ex-USS *Tacoma* (PF-3)	8 Oct 51
Nae Tong (PF-65)	ex-USS *Hoquiam* (PF-5)	8 Oct 51
Imchin (PF-66)	ex-USS *Sausalito* (PF-4)	15 Nov 52

PC 461-class Submarine Chasers – 6

Bak Du San (PC-701)	ex-USS *PC-823*	26 Dec 49
Kum Kang San (PC-702)	ex-USS *PC-799*	27 May 50
Sam Kak San (PC-703)	ex-USS *PC-802*	27 May 50
Chiri San (PC-704)	ex-USS *PC-810*	27 May 50
Han Ra San (PC-705)	ex-USS *PC-485*	3 May 52
Myo Hang San (PC-706)	ex-USS *PC-600*	3 May 52

PCS 1376-class Patrol Craft Sweepers – 5

Su Seong (PCS-201)	ex-USS *PCS-1426*	31 Aug 52
Geum Seong (PCS-202)	ex-USS *PCS-1445*	31 Aug 52
Mok Seong (PCS-203)	ex-USS *PCS-1446*	31 Aug 52
unknown	ex-USS *PCS-1442*	
Hwa Seong (PCS-205)	ex-USS *PCS-1448*	31 Aug 52

80' *Elco*-class Motor Torpedo Boats – 4

Galmaegi (PT-23)	ex-USS *PT-613*	7 Feb 52
Gireogo (PT-25)	ex-USS *PT-616*	7 Feb 52
Olbbaemi (PT-26)	ex-USS *PT-619*	7 Feb 52
Jebi (PT-27)	ex-USS *PT-620*	7 Feb 52

Gunboats – 2

Pongnoe (21)		1 Dec 49
Heukjohwan (22)	*Kuroshiomaru*	8 Apr 50

Patrol Gunboats – 2

Chungmugong I (PG-313)	Incomplete Japanese rescue ship	7 Feb 47
Chungmugong II (PG-313)		31 Aug 51

Tank Landing Ships – 5

Cheonan (LST-801)	ex-USS *LST-659*	1 Jul 49
Cheolong (LST-802)	ex-USS *LST-608*	18 Sep 51
Andong (LST-803)	ex-USS *LST-491*	15 Sep 51
Cheonbo (LST-805)	ex-USS *LST-595*	3 Sep 52
Yongbi (LST-806)	ex-USS *LST-388*	3 Sep 52

LCS(L)-3 class Large Landing Support Ships – 4

Yeongheungman (LSSL-107)	ex-USS LCS(L)(3)-77/LSSL-77	3 May 52
Ganghwaman (LSSL-108)	ex-USS LCS(L)(3)-91/LSSL-91	3 May 52
Boseongman (LSSL-109)	ex-USS LCS(L)(3)-54/LSSL-54	10 Jan 53
Yeongilman (LSSL-110)	ex-USS LCS(L)(3)-86/LSSL-86	10 Jan 53

LCI 351-class Infantry Landing Craft – 6

Seoul (LCI-101)	ex-USS *LCI(M)-594*	29 Oct 46
Jinju (LCI-102)	ex-USS *LCI(G)-516*	11 Nov 46
Chuncheon (LCI-103)	ex-USS *LCI(L)-773*	3 Jan 47
Cheongju (LCI-104)	ex-USS *LCI(G)-453*	3 Jan 47
Cheongjin (LCI-105)	ex-USS *LCI(M)-1056*	22 Jan 47
Jinnampo (LCI-106)	ex-USS *LCI(G)-442*	22 Jan 47

Yard Minesweepers YMS / British Yard Minesweepers BYMS – 19

Kang Jim (YMS-501)	ex-USS *YMS-354*	28 Apr 47
Kyong Chu (YMS-502)	ex-USS *YMS-358*	28 Apr 47
Kwang Chu (YMS-503)	ex-USS *YMS-413*	28 Apr 47
Gaeseong (YMS-504)	ex-HMS *BYMS-2006*	2 Jun 47
Kim Hae (YMS-505)	ex-USS YMS-356	2 Jun 47
Ganggye (YMS-506)	ex-USS *YMS-392*	2 Jun 47
Kang Nung (YMS-507)	ex-USS *YMS-463*	21 Jun 47
Kang Wha (YMS-508)	ex-USS *YMS-245*	11 Nov 47
Gapyeong (YMS-509)	ex-USS *YMS-220*	11 Nov 47
Ganggyeong (YMS-510)	ex-USS *YMS-330*	11 Nov 47
Kaya San (YMS-511)	ex-USS *YMS-423*	11 Nov 47
Guwolsan (YMS-512)	ex-USS *YMS-323*	11 Nov 47
Kim Chon (YMS-513)	ex-HMS *BYMS-2258*	11 Nov 47
Kil Chu (YMS-514)	ex-HMS *BYMS-2005*	21 Sep 47
Gyeongsan (YMS-515)*	ex-HMS *BYMS-2018*	21 Sep 47
Ko Yung (YMS-515)*	ex-HMS *BYMS-2055*	25 Oct 49
Gongju (YMS-516)	ex-HMS *BYMS-2148*	20 Feb 47
Gowan (YMS-517)	ex-USS *YMS-473*	11 Nov 47
Yong Kung (YMS-518)	ex-HMS *BYMS-2008*	Jul 47

*ROKS *Gyeongsan* (YMS-515) grounded, and was damaged beyond repair in February 1948; replaced by *Ko Yung* (YMS-515)

Tug Minelayers (ex-Japanese Mine-Planting Vessels) – 11

Daejeon (JMS-301)	No. 1313	11 Nov 46
Tongyeong (JMS-302)	No. 1314	11 Nov 46
Daegu (JMS-303)	No. 1372	11 Nov 46
Taebaeksan (JMS-304)	No. 1373	11 Nov 46
Dumangang (JMS-305)	No. 1121	9 Jan 47
Danyang (JMS-306)	No. 1009	8 Apr 47
Dancheon (JMS-307)	No. 1269	3 Oct 47
Toseong (JMS-308)	*Daiichikadogawamaru*	3 Oct 47
Daedonggang (JMS-309)	No. 1008	3 Oct 47
Deokcheon (JMS-310)	No. 1217	3 Oct 47
Tongcheon (JMS-311)	No. 1216	9 Jan 47

Light Cargo Ships (ex-U.S. Army Freight Supply Ships) – 5

Busan (901)	ex *FS-162*	1 Jul 49
Incheon (902)	ex *FS-198*	18 Sep 51
Wonsan (903)	ex *FS-254*	18 Sep 51
Jinnampo (905)	ex *FS-356*	27 Sep 51
Seongjin (906)	ex *FS-285*	3 Sep 52

1 Oiler, 1 Fuel Oil Barge, and 1 Large Tug – 3

Cheonji (AO-51)	*Hassel*	17 Sep 53
Guryong	USS *YO-118*	24 Dec 46
Inwang	U.S. Army *LT-134*	5 Jul 50

Appendix B: Battle Stars earned by USN Combatant Ships

211 USN Surface Combatant Ships

4 Battleships (12 Battle Stars)

Iowa (BB-61) ★★
Missouri (BB-63) ★★★★★
New Jersey (BB-62) ★★★★
Wisconsin (BB-64) ★

9 Cruisers (47 Battle Stars)

Bremerton (CA-130) ★★
Helena (CA-75) ★★★★
Juneau (CLAA-119) ★★★★★
Los Angeles (CA-135) ★★★★★
Manchester (CL-83) ★★★★★★★★★★
Rochester (CA-124) ★★★★★★
Saint Paul (CA-73) ★★★★★★★★
Toledo (CA-133) ★★★★★★
Worcester (CL-144) ★★

172 Destroyers (628 Battle Stars)
NUC – Task Element 90.62 (*Gurke, Lyman K. Swenson, Henderson, Dehaven, Mansfield, Collett*) – 15 September 1950

Agerholm (DD-826) ★★★★
Richard B. Anderson (DD-786) ★★★★
Barton (DD-722) ★★
Bausell (DD-845) ★★★
Brinkley Bass (DD-987) ★★★★★★★
Battsell (DD-945) ★★★
Beatty (DD-756) ★★
Fred T. Berry (DDE-858) ★★
Black (DD-666) ★★
Blue (DD-744) ★★★★★★
John A. Bole (DD-755) ★★★★★★★
Borie (DD-704) ★★★★
Boyd (DD-544) ★★★★★
Bradford (DD-545) ★★★★★★
Bristol (DD-857) ★★
Clarence K. Bronson (DD-668) ★
Brown (DD-546) ★★★★★
George K. Mackenzie ★★★★★★★★ (DD-836)
Maddox (DD-731) ★★★★★★
Mansfield (DD-728) ★★★★★★
Marsh (DE-699) ★★★★
Marshall (DD-676) ★★★★
Leonard F. Mason (DD-852) ★★★★★★★★
Massey (DD-778) ★★★★
McCaffery (DDE-860) ★★
McCord (DD-534) ★★
McDermut (DD-677) ★★★★★
McGinty (DE-365) ★★★
McGowan (DD-678) ★★
McKean (DDR-784) ★
McNair (DD-679) ★★
Miller (DD-535) ★★
Moale (DD-693) ★
Samuel N. Moore (DD-747) ★★★★★★

Brush (DD-745) ★★★★★
Buck (DD-761) ★★★★★★
Caperton (DD-650) ★
Carpenter (DDE-825) ★★★★★
Theodore E. Chandler ★★★★★★★★★★ (DD-717)
Chauncey (DD-667) ★★
Chevalier (DDR-805) ★★★★★★★★★★
Colahan (DD-658) ★★★★★
Collett (DD-730) ★★★★★★
Conway (DDE-507) ★★
Cony (DDE-508) ★★
Cotten (DD-669) ★
Cowell (DD-547) ★★
John R. Craig (DD-885) ★★★★
Alfred A. Cunningham ★★★★★★★ (DD-752)
Currier (DE-700) ★
Cushing (DD-797) ★★
Daly (DD-519) ★
Dashiell (DD-659) ★
Dehaven (DD-727) ★★★★★★
Dortch (DD-670) ★
Duncan (DDR-874) ★★★★★★★
Edmonds (DE-406) ★★
English (DD-696) ★★★★
Epperson (DDE-719) ★★★★★
Erben (DD-631) ★★★
Frank E. Evans (DD-754) ★★★★★★
Eversole (DD-789) ★★★★★★★
Fechteler (DD-870) ★★★★★
Fiske (DDR-842) ★★
Fletcher (DDE-445) ★★★★★
Forrest Royal (DD-872) ★★★★
Foss (DE-59) ★
Douglas H. Fox (DD-779) ★★
Gatling (DD-671) ★
Gregory (DD-802) ★★★★
Gurke (DD-783) ★★★★★★

Ulvert M. Moore (DE-442) ★★★★
Douglas A. Munro (DE-422) ★★★
Naifeh (DE-352) ★★★
Nicholas (DDE-449) ★★★★★
Norris (DDE-859) ★★
O'Bannon (DDE-450) ★★★
O'Brien (DD-725) ★★★★★★
Orleck (DD-886) ★★★★
Owen (DD-536) ★★
James C. Owens (DD-776) ★★
Ozbourn (DD-846) ★★★★★
Floyd B. Parks (DD-884) ★★★★
Perkins (DDR-877) ★★★★
Philip (DDE-498) ★★★★★★
Picking (DD-635) ★
John R. Pierce (DD-753) ★
Porter (DD-800) ★
Porterfield (DD-682) ★★★★
Stephen Potter (DD-538) ★
Halsey Powell (DD-686) ★★★★
Preston (DD-795) ★
Prichett (DD-561) ★★
Purdy (DD-734) ★★★
Radford (DDE-446) ★★★★★
Renshaw (DDE-499) ★★★★★
McCoy Reynolds (DE-440) ★
Rogers (DDR-876) ★★★★★
Rooks (DD-804) ★★
Rowan (DD-782) ★★★★
Rupertus (DD-851) ★★★★★★★★
William R. Rush (DDR-714) ★★
William Seiverling (DE-441) ★★★★★★
Shelton (DD-790) ★★★★★★
Shields (DD-596) ★★★★
Silverstein (DE-534) ★★★
Ernest G. Small (DDR-838) ★★★★
Smalley (DD-50) ★

Hailey (DD-556) ★★
Hamner (DD-718) ★★★★★★★
Lewis Hancock (DD-675) ★★
Hank (DD-702) ★★★★
Hanna (DE-449) ★★★★★
Hanson (DDR-832) ★★★★★★★
Hawkins (DDR-873) ★★
Henderson (DD-785) ★★★★
Hickox (DD-673) ★★
Higbee (DDR-806) ★★★★★★
Hollister (DD-788) ★★★★★
John Hood (DD-655) ★
Hopewell (DD-681) ★★★★
Harry E. Hubbard (DD-748) ★★★★★
Hyman (DD-732) ★★

Ingraham (DD-694) ★
Irwin (DD-794) ★
Arnold J. Isbell (DD-869) ★★★★★★

Jarvis (DD-799) ★
Jenkins (DDE-447) ★
Joseph P. Kennedy (DD-850) ★★
Keppler (DDE-765) ★★
James E. Keyes (DD-787) ★★★★★★★
Kidd (DD-661) ★★★★
Kimberly (DD-521) ★
Frank Knox (DDR-742) ★★★★★
Laffey (DD-724) ★★
Laws (DD-558) ★★
Lewis (DE-535) ★
Wallace L. Lind (DD-703) ★★★★
Lofberg (DD-759) ★★★★★★★
Lowry (DD-770) ★★

Soley (DD-707) ★
Southerland (DDR-743) ★★★★★★★★
Charles S. Sperry (DD-697) ★★★★
Sproston (DDE-577) ★
Stembel (DD-644) ★★
Stickell (DD-888) ★★★★★★
Stormes (DD-780) ★★
Strong (DD-758) ★
Allen M. Sumner (DD-692) ★
Lyman K. Swenson (DD-729) ★★★★★★
Taussig (DD-746) ★★★★★★★★
Taylor (DDE-468) ★★
The Sullivans (DD-537) ★★
Herbert J. Thomas (DDR-833) ★★★★★★★
John W. Thomason ★★★★★★★
(DD-760)
Tingey (DD-539) ★★★★★
Trathen (DD-530) ★★
Henry W. Tucker ★★★★★★★
(DDR-875)
Twining (DD-540) ★★★★★
Uhlmann (DD-687) ★★★★★
Vammen (DE-644) ★
Van Valkenburgh (DD-656) ★
Walke (DD-723) ★★★★
Walker (DDE-517) ★★
Waller (DDE-466) ★★
Walton (DE-361) ★★
Wedderburn (DD-684) ★★★★
Whitehurst (DE-634) ★★★
Wiltsie (DD-716) ★★★★★★★★★
Wiseman (DE-667) ★★★★★★
Yarnall (DD-541) ★★★★★
Zellars (DD-777) ★★★★

4 Destroyer Minesweepers (DMS) (27 Battle Stars)
NUC - *Carmick* (DMS 33) and *Thompson* (DMS 38) – Port of Chinnampo

Carmick (DMS-33) ★★★★★★	*Endicott* (DMS-35) ★★★★★★★★★
Doyle (DMS-34) ★★★★★★	*Thompson* (DMS-38) ★★★★★★★★

5 High-Speed Transports (APD) – Converted Destroyers (26 Battle Stars)

Horace A. Bass (APD-124) ★★★★★★	*Wantuck* (APD-125) ★★★★★
Begor (APD-127) ★★★★★★	*Weiss* (APD-135) ★★★
Alex Diachenko (APD-123) ★★★★★★	

13 Patrol Frigates (59 Battle Stars)

Albuquerque (PF-7) ★★	*Glendale* (PF-36) ★★★★
Bayonne (PF-21) ★★★★★★	*Gloucester* (PF-22) ★★★★★★★★
Bisbee (PF-46) ★★★★	*Hoquiam* (PF-5) ★★★★★
Burlington (PF-51) ★★★★★	*Newport* (PF-27) ★★★★
Evansville (PF-70) ★★★★★★	*Sausalito* (PF-4) ★★★★★★★
Everett (PF-8) ★★★★	*Tacoma* (PF-3) ★★★
Gallup (PF-47) ★★★	

4 Unnamed Patrol Escorts (15 Battle Stars)

PCE(C)-882 ★★★★	*PCE(C)-896* ★★★★★★
PCE(C)-886 ★★	*PCE(C)-898* ★★★

NUC: Navy Unit Commendation

Appendix C: Patrol Frigates that served in the Korean War

Kaiser Cargo Inc., Richmond, California

Ship	Comm	Decom	Disposition
USS *Tacoma* (PF-3) Soviet EK-12; returned 16 Oct 49 at Yokosuka	6 Nov 43 1 Dec 50	16 Aug 45 at CH early Oct 51	To ROK 9 Oct 51, *Taedong* (PF-63)
USS *Sausalito* (PF-4) Soviet EK-13; returned 1 Nov 49 at Yokosuka	4 Mar 44 15 Sep 50	16 Aug 45 at CH 9 Jun 52	To ROK 4 Sep 52, *Imchin* (PF-66)
USS *Hoquiam* (PF-5) Soviet EK-14; returned 1 Nov 49 at Yokosuka	8 May 44 27 Sep 50	16 Aug 45 at CH 5 Oct 51	To ROK, *Nae Tong* (PF-65)
USS *Albuquerque* (PF-7) Soviet EK-16; returned 15 Nov 49 at Yokosuka	20 Dec 43 30 Oct 50	16 Aug 45 at CH 28 Feb 53	To Japan 30 Nov 53, *Tochi* (PF 296)
USS *Everett* (PF-8) Soviet EK-17; returned 15 Nov 49 at Yokosuka	22 Jan 44 26 Jul 50	16 Aug 45 at CH 10 May 53	To Japan, *Kiri* (PF-291)

American Shipbuilding Co., Cleveland, Ohio

Ship	Comm	Decom	Disposition
USS *Bayonne* (PF-21) Soviet EK-24; returned 14 Nov 49 at Yokosuka	14 Feb 45 28 Jul 50	2 Sep 45 at CH 31 Jan 53	To Japan 31 Jan 53, *Buna* (PF-294)

Walter Butler Shipbuilding Co., Superior, Wisconsin

Ship	Comm	Decom	Disposition
USS *Gloucester* (PF-22) Soviet EK-26; returned 31 Oct 49 at Yokosuka	10 Dec 43 11 Oct 50	3 Sep 45 at CH 15 Sep 52	To Japan 1 Oct 53, *Tsuge* (PF-292)
USS *Newport* (PF-27) Soviet EK-28; returned 4 Nov 49 at Yokosuka	8 Sep 44 27 Jul 50	9 Sep 45 at CH 30 Apr 52	To Japan 1 Oct 53, *Kaede* (PF-13)

Consolidated Steel Corp, Ltd., Wilmington, California

Ship	Comm	Decom	Disposition
USS *Glendale* (PF-36) Soviet EK-6; returned 16 Nov 49 at Yokosuka	1 Oct 43 11 Oct 50	12 Jun 45 at CH 29 Oct 951	To Thailand, HTMS *Tachin* (PF-1)
USS *Burlington* (PF-51) Soviet EK-21; returned 14 Nov 49 at Yokosuka	3 Apr 44 5 Jan 51	26 Aug 45 at CH 15 Sep 52	To Colombia 26 June 53, ARC *Almirante Brion* (F-14)

Consolidated Steel Corp, Ltd., Los Angeles, California

USS *Bisbee* (PF-46) Soviet EK-17; returned 1 Nov 49 at Yokosuka	15 Feb 44 18 Oct 50	26 Aug 45 at CH 20 Oct 51	To Colombia 13 Feb 52, ARC *Capitan Tono* (F-6)
USS *Gallup* (PF-47) Soviet EK-22; returned 14 Nov 49 at Yokosuka	29 Feb 44 18 Oct 50	26 Aug 45 at CH 29 Oct 51	To Thailand, HTMS *Prasae* (PF-2)

Leathem D. Smith Shipyard, Sturgeon Bay, Wisconsin

USS *Evansville* (PF-70) Soviet EK-30; returned 17 Feb 50 at Yokosuka[1]	4 Dec 44 29 Jul 50	9 Sep 45 at CH 28 Feb 53	To Japan 31 Oct 53, *Keyaki* (PF-295)

CH: Transferred to Soviets at Cold Harbor, Alaska

Appendix D: U.S. Navy Ships Sunk (5) and Damaged in Action (87 Incidents)

Ships Sunk

Ship	Action/Date
Magpie (AMS-25)	blew up after striking a mine, 21 missing in action and 12 survivors, 29 September 1950
Pirate (AM-275)	sunk after striking a mine at Wonsan, North Korea, 12 October 1950.
Pledge (AM-277)	sunk after striking a mine at Wonsan, North Korea, 12 October 1950
Partridge (AMS-31)	sunk after striking a mine, 8 killed, 6 seriously wounded, 2 February 1951
Sarsi (ATF-111)	sunk after striking a mine at Hungnam, North Korea, 2 killed, 27 August 1952

Ships Damaged

Ship	Action/Date
Collett (DD-730)	damaged by 7 hits with 5 wounded, 13 September 1950
Gurke (DD-783)	minor damage from 3 hits, no casualties, 13 September 1950
Lyman K. Swenson (DD-729)	2 near misses caused 1 killed and 1 wounded, 13 September 1950
Brush (DD-745)	damaged after striking a mine at Tanchon, North Korea, 9 killed and 10 wounded, 26 September 1950
Mansfield (DD-728)	damaged after striking a mine, 5 missing and 48 wounded, 30 September 1950
Charles S. Sperry (DD-697)	damaged by 3 hits from a shore battery at Songjin, North Korea, 23 December 1950
Ozbourn (DD-846)	damaged after being hit by a shore battery at Wonsan, North Korea, 2 casualties, 23 December 1950
Walke (DD-723)	extensively damaged after striking a mine off the east coast of Korea, 61 casualties, 12 June 1951
Thompson (DMS-38)	extensively damaged after being hit by a shore battery at Songjin, North Korea, 3 killed and 4 wounded, 14 June 1951
Hoquiam (PF-5)	slightly damaged after being hit by a shore battery at Songjin, North Korea, 1 casualty, 7 May 1951
New Jersey (BB-62)	slightly damaged after being hit by a shore battery at Wonsan, North Korea, 4 casualties, 20 May 1951
Brinkley Bass (DD-887)	minor damage after being hit by a shore battery at Wonsan, North Korea, 8 casualties, 22 May 1951
Frank E. Evans (DD-754)	slightly damaged after being hit by a shore battery at Wonsan, North Korea, 4 casualties, 18 June 1951

Henry W. Tucker (DDR-875)	superficial damage after being hit by a shore battery at Wonsan, North Korea, 28 June 1951
Everett (PF-8)	minor damage after being hit by a shore battery at Wonsan, North Korea, 8 casualties, 3 July 1951
Helena (CA-75)	minor damage after being hit by a shore battery at Wonsan, North Korea, 2 casualties, 31 July 1951
Dextrous (AM-341)	superficial damage after being hit by a shore battery at Wonsan, North Korea, 1 killed and 3 wounded, 11 August 1951
William Seiverling (DE-441)	fireroom flooded after being hit by a shore battery at Wonsan, North Korea, no casualties, 8 September 1951
Heron (AMS-18)	superficial damage after being hit by a shore battery at Wonsan, North Korea, no casualties, 10 September 1951
Redstart (AM-378)	minor damage after being hit by a shore battery at Wonsan, North Korea, no casualties, 10 September 1951
Firecrest (AMS-10)	slight damage after being hit by a shore battery at Hungnam, North Korea, no casualties, 5 October 1951
Ernest G. Small (DDR-838)	extensive damage after striking a mine off the East coast of North Korea, 27 casualties, 7 October 1951
Renshaw (DDE-499)	slight damage after being hit by a shore battery at Songjin, North Korea, 1 casualty, 11 October 1951
Ulvert M. Moore (DE-442)	moderate damage after being hit by a shore battery at Hungnam, North Korea, 3 casualties, 17 October 1951
Helena (CA-75)	slight damage after being hit by a shore battery at Hungnam, North Korea, 4 casualties, 23 October 1951
Osprey (AMS-28)	considerable damage after being hit by a shore battery at Wonsan, North Korea, 1 casualty, 29 October 1951
Gloucester (PF-22)	light damage after being hit by a shore battery at Hongwon, North Korea, 12 casualties, 11 November 1951
Hyman (DD-732)	minor damage after being hit by a shore battery at Wonsan, North Korea, no casualties, 23 November 1951
LST-611	superficial damage after being hit by a shore battery, no casualties, 22 December 1951
Dextrous (AM-341)	minor damage after being hit by a shore battery at Wonsan, North Korea, 3 casualties, 11 January 1952
Porterfield (DD-682)	minor damage after being hit by a shore battery at Sok-to, North Korea, no casualties, 3 February 1952
Endicott (DMS-35)	minor damage after 2 hits from a shore battery at Songjin, North Korea, no casualties, 4 February 1952
Rowan (DD-782)	minor damage after 1 hit from a shore battery at Hungnam, North Korea, no casualties, 22 February 1952
Shelton (DD-790)	moderate damage after 3 hits from a shore battery at Songjin, North Korea, 15 casualties, 22 February 1952
Henderson (DD-785)	minor damage after being hit by a shore battery at Hungnam, North Korea, no casualties, 23 February 1952
Wisconsin (BB-64)	insignificant damage after 1 hit from a shore battery at Songjin, North Korea, 3 casualties, 16 March 1952
Brinkley Bass (DD-887)	moderate damage after 1 hit from a shore battery at Wonsan, North Korea, 5 casualties, 24 March 1952

U.S. Navy Ships Sunk and Damaged in Action

Endicott (DMS-35)	insignificant damage after being hit by a shore battery at Chongjin, North Korea, no casualties, 7 April 1952
Endicott (DMS-35)	minor damage after 1 hit from a shore battery at Songjin, North Korea, no casualties, 19 April 1952
Osprey (AMS-28)	minor damage after 1 hit from a shore battery at Songjin, North Korea, no casualties, 24 April 1952
Cabildo (LSD-16)	minor damage after 1 hit from a shore battery at Wonsan, North Korea, 2 casualties, 26 April 1952
Laffey (DD-724)	superficial damage after being hit by a shore battery at Wonsan, North Korea, no casualties, 30 April 1952
Maddox (DD-731)	superficial damage after being hit by a shore battery at Wonsan, North Korea, no casualties, 30 April 1952
Leonard F. Mason (DD-852)	superficial damage after being hit by a shore battery at Wonsan, North Korea, no casualties, 2 May 1952
James C. Owens (DD-776)	considerable damage after 6 hits from a shore battery at Songjin, North Korea, 10 casualties, 7 May 1952
Herbert J. Thomas (DDR-833)	superficial damage after 1 hit from a shore battery at Wonsan, North Korea, no casualties, 12 May 1952
Douglas H. Fox (DD-779)	minor damage after 1 hit from a shore battery at Hungnam, North Korea, 2 casualties, 14 May 1952
Cabildo (LSD-16)	superficial damage after being hit by a shore battery at Wonsan, North Korea, 2 casualties, 25 May 1952
Swallow (AMS-26)	slight damage after 3 hits from a shore battery at Songjin, North Korea, no casualties, 25 May 1952
Murrelet (AM-372)	slight damage after being hit by a shore battery at Songjin, North Korea, no casualties, 26 May 1952
Firecrest (AMS-10)	minor damage after hits from machine gun mounts. No casualties, 30 May 1952
Buck (DD-761)	motor launch damaged after being hit by a shore battery at Kojo, North Korea, 2 casualties, 13 June 1952
Orleck (DD-886)	minor damage 1 hit after receiving 50 rounds of 75 mm, 4 casualties, 13 June 1952
Southerland (DDR-743)	minor damage after 4 hits from shore batteries, 8 casualties, 14 July 1952
John R. Pierce (DD-753)	moderate damage after 7 hits from a shore battery at Tanchon, North Korea, 10 casualties, 6 August 1952
Barton (DD-722)	minor damage after 1 hit from a shore battery at Wonsan, North Korea, 2 casualties, 10 August 1952
Grapple (ARS-7)	minor damage after 1 hit below the waterline from a shore battery at Wonsan, North Korea, no casualties, 12 August 1952
Thompson (DMS-38)	minor damage in the vicinity of the bridge after an air burst and near misses from a shore battery at Songjin, North Korea, 13 casualties, 20 August 1952
Competent (AM-316)	superficial damage and lost sweep gear after a shrapnel near miss from a shore battery at Pkg. 4-5 (package 4-5 refers to a designated train target area), no casualties, 27 August 1952
McDermut (DD-677)	superficial damage after receiving 60 rounds at 3,700 yards while at Pkg 4-5, no casualties, 27 August 1952

Appendix D

Agerholm (DD-826)	superficial damage after being hit by a shore battery at the Kangsong, North Korea area bombline, 1 casualty, 1 September 1952
Frank E. Evans (DD-754)	slight damage from near misses, after receiving 69 rounds, from a shore battery at Tanchon, North Korea, no casualties, 8 September 1952
Barton (DD-722)	major damage after striking a mine 90 miles east of Wonsan, North Korea, 11 casualties, 16 September 1952
Alfred A. Cunningham (DD-752)	moderate damage from 5 hits and 7 air bursts. Received 150 rounds of 105 mm from 3 guns. First round was a direct hit at an initial range of 3,500 yards. 8 casualties, 19 September 1952
Perkins (DDR-877)	superficial damage after being straddled by 5 rounds, from a shore battery at range of 5,000 yards, at Kojo, North Korea. The ship was sprayed with shrapnel from 2 near misses, 18 casualties, 13 October 1952
Osprey (AMS-28)	minor damage after being hit by a shore battery at Kojo, North Korea, 4 casualties, 14 October 1952
Lewis (DE-535)	moderate damage from 2 hits after receiving 50 rounds from 4-6 guns at Wonsan, North Korea, 8 casualties, 21 October 1952
Mansfield (DD-728)	minor shrapnel damage after receiving 40 rounds from 4 guns. The suspected radar-controlled guns straddled the ship at a range of 4,300-8,000 yards. No casualties, 28 October 1952
Uhlmann (DD-687)	minor damage from 3 hits after receiving 160 rounds from a shore battery, 13 casualties, 3 November 1952
Kite (AMS-22)	1 small boat destroyed by a shore battery at Wonsan, North Korea, 5 casualties, 19 November 1952
Thompson (DMS-38)	minor damage from 1 hit after receiving 89 rounds from a shore battery at Wonsan, North Korea, 1 casualty, 20 November 1952
Hanna (DE-449)	moderate damage from 1 hit after receiving 60 rounds from a shore battery at Songjin, North Korea, 1 casualty, 24 November 1952
Halsey Powell (DD-686)	whaleboat damaged after being hit by a shore battery at Hwa-do, North Korea, 2 casualties, 6 February 1953
Gull (AMS-16)	minor damage from 1 hit after receiving 60 rounds at a range of 5,400-10,000 yards while at Pkg 2, 2 casualties, 16 March 1953
Taussig (DD-746)	slight damage from 1 hit after receiving 45 rounds at a range of 6,400-10,000 yards, 1 casualty, 17 March 1953
Los Angeles (CA-135)	slight damage from 1 hit after receiving 40 rounds of 105 mm at Wonsan, North Korea, no casualties, 27 March 1953
Los Angeles (CA-135)	minor damage after 1 hit from a shore battery at Wonsan, North Korea, 13 casualties, 2 April 1953
Maddox (DD-731)	slight damage from one 76 mm hit after receiving 209 rounds of heavy fire from a shore battery at Wonsan, North Korea, 3 casualties, 16 April 1953

U.S. Navy Ships Sunk and Damaged in Action

James E. Kyes (DD-787)	slight damage from 1 hit after receiving 60 rounds of 155 mm at a range of 8,000-12,000 yards from a shore battery near Wonsan, North Korea, 9 casualties, 19 April 1953
Maddox (DD-731)	moderate damage from 1 hit from a shore battery at Hodo Pando. The ship received 186 rounds of 105 mm and several near misses from 4 guns. No casualties, 2 May 1953
Owen (DD-536)	minor damage from 1 hit from a shore battery at Hodo Pando, North Korea. The ship received 100 rounds of 105 mm with 1 near miss and several straddles from 4 guns. No casualties, 2 May 1953
Bremerton (CA-130)	superficial damage after 1 near miss from a shore battery at Wonsan, North Korea. The ship received 18 rounds of 76 mm - 135 mm, 2 casualties, 5 May 1953
Samuel N. Moore (DD-747)	superficial damage from 1 hit from a shore battery at Wonsan, North Korea. The ship received 60 rounds of 90 mm, no casualties, 8 May 1953
Brush (DD-745)	minor damage after 1 hit from a shore battery at Wonsan, North Korea. The ship received 20 rounds of 76 mm, 9 casualties, 15 May 1953
Swift (AM-122)	superficial damage from 1 hit from a shore battery at Yang-do, North Korea. The ship received 30 rounds of 76 mm, 1 casualty, 29 May 1953
Clarion River (LSMR-409)	minor damage after 2 hits from a shore battery at Walsa-ri, North Korea. The ship received 30 rounds of 76 mm, 5 casualties, 4 June 1953
Wiltsie (DD-716)	superficial damage after 1 hit from a shore battery at Wonsan, North Korea. The ship received 35 rounds of 76 mm with several air bursts, no casualties, 11 June 1953
Henderson (DD-785)	superficial damage after being hit by a shore battery at Wonsan, North Korea, 17 June 1953
Irwin (DD-794)	minor damage from 1 hit after receiving 90 rounds near Wonsan, North Korea, 5 casualties, 18 June 1953
Rowan (DD-782)	moderate damage from 5 hits after receiving 45 rounds of 76-155 mm, at 7,500 yards, near Wonsan, North Korea, 9 casualties, 18 June 1953
Gurke (DD-783)	slight damage from 2 hits and shrapnel from 5 near misses after receiving 150 rounds of 76-90 mm, at 6,000 to 11,000 yards, near Songjin, North Korea, 3 casualties, 25 June 1953
Manchester (CL-83)	superficial damage after near misses during a 30 minute gun duel with a shore battery at Wonsan, North Korea, no casualties, 30 June 1953
John W. Thomason (DD-760)	minor shrapnel damage after near misses from 150 rounds of 107 mm from a shore battery at Hodo Pando, North Korea, 7 July 1953
Irwin (DD-794)	minor damage after 80 rounds of 76 mm air bursts close aboard from a shore battery at Pkg 2, 5 casualties, 8 July 1953
Saint Paul (CA-73)	severe underwater damage after one 76-90 mm hit from a shore battery at Wonsan, no casualties, 11 July 1953[2]

Appendix E: Allied Ships (less USN and ROKN) Deployments

ARC: Armada de la Republica de Colombia
FS: French Ship
HMS: His/Her Majesty's Ship
HMAS: His/Her Majesty's Australian Ship
HMCS: His/Her Majesty's Canadian Ship
HMNZS: His/Her Majesty's New Zealand Ship
HTMS: His Thai Majesty's Ship
HNLMS: His/Her Netherlands Majesty's Ship

The dates in the table generally reflect when a ship arrived in theater, for many this was considered Sasebo, and when they departed it, or were relieved of their duties by the ship that replaced them. Entries for ships of the Royal Navy reflect the month and year they began a patrol in Korea waters, and when they completed it. These periods do not include time spent in port, for provisioning, maintenance, or crew rest.

1950

Month	Arrived in Theater	Departed Theater/Relieved
June	HMS *Alacrity* (F60)	
	HMS *Belfast* (C35)	
	HMS *Black Swan* (F57)	
	HMS *Consort* (D76)	
	HMS *Cossack* (D57)	
	HMS *Hart* (F58)	
	HMS *Jamaica* (C44)	
	10 - HMAS *Bataan*	
	27 - HMAS *Shoalhaven*	
July	HMS *Charity* (D29)	
	HMS *Cockade* (D34)	
	HMS *Comus* (D43)	
	HMS *Kenya* (C14)	
	3 - HMNZS *Pukaki*	
	3 - HMNZS *Tutira*	
	16 - HNLMS *Evertsen*	
	29 - FS *La Grandiere*	
	30 - HMCS *Athabaskan*	
	30 - HMCS *Cayuga*	
	30 - HMCS *Sioux*	

Month	Arrived in Theater	Departed Theater/Relieved
August	HMS *Alert* (F647)	HMS *Alacrity* (F60)
	HMS *Ceylon* (C30)	HMS *Belfast* (C35)
	HMS *Mounts Bay* (F627)	HMS *Black Swan* (F57)
	HMS *Whitesand Bay* (F633)	HMS *Comus* (D43)
	14 - HMAS *Warramunga*	HMS *Hart* (F58)
September	HMS *Concord* (D63)	22 - HMAS *Shoalhaven*
October	HMS *Constant* (D71)	HMS *Alert* (F647)
	HMS *Morecambe Bay* (F624)	HMS *Jamaica* (C44)
	7 - HMNZS *Rotoiti*	
November	HMS *Cardigan Bay* (F630)	HMS *Cockade* (D34)
	7 - HTMS *Bangpakong*	HMS *Mounts Bay* (F627)
	7 - HTMS *Prasae*	25 - FS *La Grandiere*
December	HMS *Mounts Bay* (F627)	HMS *Whitesand Bay* (F633)
	HMS *St. Brides Bay* (F600)	3 – HMNZS *Pukaki*

1951

Month	Arrived in Theater	Departed Theater/Relieved
January	HMS *Belfast* (C35)	HMS *Cardigan Bay* (F630)
	HMS *Comus* (D43)	HMS *Charity* (D29)
	14 - HMCS *Nootka*	HMS *Concord* (D63)
		HMS *Morecambe Bay* (F624)
		HMS *Mounts Bay* (F627)
		HMS *St. Brides Bay* (F600)
		7 - HTMS *Prasae*
		15 - HMCS *Sioux*
February	HMS *Alacrity* (F60)	
	HMS *Amethyst* (F116)	
	HMS *Black Swan* (F57)	
	HMS *Hart* (F58)	
March	HMS *Cockade* (D34)	HMS *Constant* (D71)
	2 - HMNZS *Hawea*	HMS *Hart* (F58)
	15 - HMCS *Huron*	16 - HMCS *Cayuga*
April	HMS *Concord* (D63)	HMS *Consort* (D76)
	18 - HNLMS *Van Galen*	
	30 - HMCS *Sioux*	
May	5 - ARC *Almirante Padilla*	HMS *Concord* (D63)
	9 - HMAS *Murchison*	HMNZS *Tutira*
		3 - HMCS *Athabaskan*
June	HMS *Cardigan Bay* (F630)	HMS *Alacrity* (F60)
	HMS *Consort* (D76)	HMS *Amethyst* (F116)
	HMS *Constant* (D71)	HMS *Black Swan* (F57)
	HMS *Morecambe Bay* (F624)	HMS *Comus* (D43)
	HMS *Mounts Bay* (F627)	6 - HMAS *Bataan*
	HMS *Whitesand Bay* (F633)	
July	HMS *Charity* (D29)	HMS *Constant* (D71)
	20 - HMCS *Cayuga*	HMS *Whitesand Bay* (F633)
		20 - HMCS *Nootka*

Allied Ships (less USN and ROKN) Korean Deployments 251

August	HMS *Concord* (D63)	HMS *Cockade* (D34)
	HMS *St. Brides Bay* (F600)	HMS *Kenya* (C14)
	6 - HMAS *Anzac*	29 - HMAS *Warramunga*
	29 - HMNZS *Taupo*	
	31 - HMAS *Tobruk*	
September	HMS *Alert* (F647)	HMS *Cardigan Bay* (F630)
	HMS *Amethyst* (F116)	HMS *Charity* (D29)
	HMS *Black Swan* (F57)	HMS *Consort* (D76)
	HMS *Comus* (D43)	HMS *Belfast* (C35)
	1 - HMCS *Athabaskan*	HMS *Morecambe Bay* (F624)
		HMS *Mounts Bay* (F627)
October	HMS *Cockade* (D34)	HMS *Alert* (F647)
	HMS *Whitesand Bay* (F633)	HMS *Cossack* (D57)
		17 - HMAS *Anzac*
November	HMS *Constant* (D71)	HMS *Concord* (D63)
		HMS *Black Swan* (F57)
		5 - HNLMS *Evertsen*
		21 - HMNZS *Rotoiti*
December	HMS *Alacrity* (F60)	HMS *Cockade* (D34)
	HMS *Charity* (D29)	HMS *Comus* (D43)
	HMS *Mounts Bay* (F627)	HMS *St. Brides Bay* (F600)
	29 - HTMS *Prasae II*	
	29 - HTMS *Tachin*	

1952

Month	Arrived in Theater	Departed Theater/Relieved
January	HMS *Cardigan Bay* (F630)	HMS *Amethyst* (F116)
	HMS *Cockade* (D34)	
	HMS *Concord* (D63)	
	7 - HMNZS *Rotoiti*	
	17 - HMAS *Bataan*	
	17 - HMAS *Warramunga*	
February	HMS *Cossack* (D57)	HMS *Alacrity* (F60)
	12 - ARC *Capitan Tono*	HMS *Constant* (D71)
	12 - HMCS *Nootka*	HMS *Whitesand Bay* (F633)
		12 - ARC *Almirante Padilla*
		14 - HMCS *Sioux*
		16 - HTMS *Bangpakong*
		17 - HMAS *Murchison*
		23 - HMAS *Tobruk*
March	HMS *Comus* (D43)	HMS *Charity* (D29)
	HMS *Crane* (F123)	HMS *Cockade* (D34)
	HMS *Morecambe Bay* (F624)	2 - HNLMS *Van Galen*
	2 - HMNZS *Kaniere*	8 - HMNZS *Hawea*
	2 - HNLMS *Piet Hein*	
April	HMS *Amethyst* (F116)	HMS *Cardigan Bay* (F630)
	HMS *Whitesand Bay* (F633)	HMS *Concord* (D63)
		HMS *Mounts Bay* (F627)
May	HMS *Consort* (D76)	HMS *Cossack* (D57)
		HMS *Morecambe Bay* (F624)

Month	Arrived in Theater	Departed Theater/Relieved
June	HMS *Cardigan Bay* (F630)	HMS *Crane* (F123)
	HMS *Constant* (D71)	1 - HMCS *Cayuga*
	HMS *Mounts Bay* (F627)	21 - HMCS *Athabaskan*
	12 - HMCS *Iroquois*	
	21 - HMCS *Crusader*	
July	HMS *Concord* (D63)	HMS *Amethyst* (F116)
	HMS *Newcastle* (C76)	HMS *Ceylon* (C30)
	HMS *St. Brides Bay* (F600)	HMS *Whitesand Bay* (F633)
	4 - HMAS *Condamine*	
August	HMS *Charity* (D29)	HMS *Concord* (D63)
	HMS *Crane* (F123)	HMS *Consort* (D76)
	HMS *Morecambe Bay* (F624)	8 - HMAS *Warramunga*
	4 - HMNZS *Hawea*	
September	HMS *Birmingham* (C19)	HMS *Cardigan Bay* (F630)
	HMS *Cossack* (D57)	HMS *Comus* (D43)
	6 - HMAS *Anzac*	HMS *Crane* (F123)
		25 - HMAS *Bataan*
October		HMS *St. Brides Bay* (F600)
		21 - HMNZS *Taupo*
November	HMS *Comus* (D43)	HMS *Charity* (D29)
	HMS *Consort* (D76)	HMS *Morecambe Bay* (F624)
	HMS *Crane* (F123)	HMS *Mounts Bay* (F627)
	HMS *Opossum* (F33)	9 - HMCS *Nootka*
	6 - HMCS *Haida*	26 - HMCS *Iroquois*
	26 - HMCS *Athabaskan*	
December	HMS *Cockade* (D34)	HMS *Constant* (D71)
	HMS *Sparrow* (F71)	

1953

Month	Arrived in Theater	Departed Theater/Relieved
January	18 - HNLMS *Johan Maurits van Nassau*	HMS *Cossack* (D57)
		18 - HNLMS *Piet Hein*
	27 - ARC *Almirante Brion*	27 - ARC *Capitan Tono*
February	HMS *Charity* (D29)	HMS *Cockade* (D34)
	HMS *Whitesand Bay* (F633)	HMS *Comus* (D43)
		HMS *Consort* (D76)
		HMS *Sparrow* (F71)
March	HMS *Consort* (D76)	HMS *Crane* (F123)
	HMS *Mounts Bay* (F627)	19 - HMNZS *Rotoiti*
	14 - HMAS *Culgoa*	
April	HMS *Cockade* (D34)	HMS *Charity* (D29)
	HMS *Modeste* (F42)	HMS *Opossum* (F33)
	HMS *Sparrow* (F71)	10 - HMAS *Condamine*
	HMS *St. Brides Bay* (F600)	
May	HMS *Concord* (D63)	HMS *Consort* (D76)
	HMS *Cossack* (D57)	
	HMS *Morecambe Bay* (F624)	

Allied Ships (less USN and ROKN) Korean Deployments

June	HMS *Charity* (D29)	HMS *Cossack* (D57)
	3 - HMAS *Tobruk*	HMS *Mounts Bay* (F627)
	18 - HMCS *Iroquois*	HMS *Sparrow* (F71)
		HMS *St. Brides Bay* (F600)
		12 - HMCS *Haida*
		18 - HMCS *Crusader*
		26 - HMAS *Anzac*
		26 - HMAS *Culgoa*
July	HMS *Crane* (F123)	HMS *Birmingham* (C19)
		HMS *Charity* (D29)
		HMS *Cockade* (D34)
		HMS *Concord* (D63)
		HMS *Crane* (F123)
		HMS *Modeste* (F42)
		HMS *Morecambe Bay* (F624)
		HMS *Newcastle* (C76)
		HMS *Whitesand Bay* (F633)
		18 - ARC *Almirante Brion*
		27 - HNLMS *Johan Maurits van Nassau*
August		29 - HMNZS *Hawea*
November	20 - HMCS *Crusader*	18 - HMCS *Athabaskan*
December		

1954

Month	Arrived in Theater	Departed Theater/Relieved
January	1 - HMCS *Cayuga*	1 - HMCS *Iroquois*
February	5 - HMCS *Haida*	5 - HMCS *Huron*
		12 - HMAS *Tobruk*
March		2 - HMNZS *Kaniere*
August	22 - HMCS *Iroquois*	15 - HMCS *Crusader*
September		12 - HMCS *Haida*
October	1 - HMCS *Huron*	
November		22 - HMCS *Cayuga*
December	14 - HMCS *Sioux*	26 - HMCS *Huron*
		26 - HMCS *Iroquois*

1955

Month	Arrived in Theater	Departed Theater/Relieved
January		21 - HTMS *Prasae II*
		21 - HTMS *Tachin*
September		7 - HMCS *Sioux*

Appendix F: Cruisers, Destroyers, and Frigates assigned to the Inchon Landings

Joint Task Force 7: Vice Adm. Arthur D. Struble, USN

Task Force 77 (Fast Carrier): Rear Adm. Edward C. Ewen, USN

Carrier Division 1	Rear Adm. Edward C. Ewen, USN
USS *Philippine* Sea (CV-47)	Capt. William K. Goodney, USN
Carrier Division 2	Rear Adm. John M. Hoskins, USN
USS *Valley Forge* (CV-45)	Capt. Lester K. Rice, USN
Carrier Division 5	
USS *Boxer* (CV-21)	Capt. Cameron Briggs, USN
77.1 Support Group	Capt. Harry H. Henderson, USN
USS *Worcester* (CL-144)	Capt. Harry H. Henderson, USN
77.2 Screen Group	Capt. Charles W. Parker, USN
Destroyer Division 31	Capt. Charles W. Parker, USN
USS *Shelton* (DD-790)	Comdr. Charles B. Jackson Jr., USN
USS *James E. Kyes* (DD-787)	Comdr. Fran M. Christiansen, USN
USS *Eversole* (DD-789)	Comdr. Charles E. Phillips, USN
USS *Higbee* (DDR-806)	Comdr. Elmer Moore, USN
Destroyer Division 111	Capt. Jeane R. Clark, USN
USS *Wiltsie* (DD-716)	Comdr. Carrol W. Brigham, USN
USS *Theodore E. Chandler* (DD-717)	Comdr. William J. Collum Jr., USN
USS *Hamner* (DD-718)	Comdr. Jack J. Hughes, USN
USS *Chevalier* (DDR-805)	Comdr. Blake B. Booth, USN
Destroyer Division 112	Capt. Bernard F. Roeder, USN
USS *Ozborne* (DD-846)	Comdr. Charles O. Akers, USN
USS *McKean* (DD-784)	Comdr. Harry L. Reiter Jr., USN
USS *Hollister* (DD-788)	Comdr. Hugh W. Howard, USN
USS *Frank Knox* (DDR-742)	Comdr. Sam J. Caldwell Jr., USN
Escort Squadron 1	
USS *Fletcher* (DDE-445)	Comdr. W. M. Lowry, USN
USS *Radford* (DDE-446)	Comdr. Elvin C. Ogle, USN

Task Force 90 (Attack Force): Rear Adm. James H. Doyle, USN
Task Group 90.5 (Air Support Group):
Rear Adm. Richard W. Ruble, USN

90.51 Escort Carriers	Rear Adm. Richard W. Ruble, USN
USS *Badoeng Strait* (CVE-116)	Capt. Arnold W. McKechnie, USN
USS *Sicily* (CVE-118)	Capt. John S. Thach, USN
90.52 Escort Carrier Screen	Comdr. Byron L. Gurnette, USN
USS *Hanson* (DDR-832)	Comdr. Cecil R. Welte, USN
USS *Taussig* (DD-746)	Comdr. William C. Meyer, USN
USS *George K. MacKenzie* (DD-836)	Comdr. William R. Laird Jr., USN
USS *Ernest G. Small* (DD-838)	Comdr. Franklin C. Snow, USN

Task Group 90.6 (Gunfire Support Group):
Rear Adm. John M. Higgins, USN

90.6.1 Fire Support Unit 1	Rear Adm. John M. Higgins, USN
USS *Toledo* (CA-133)	Capt. Richard F. Stout, USN
USS *Rochester* (CA-124)	Capt. Edward L. Woodyard, USN
HMS *Kenya* (C14)	Capt. P. W. Brock, RN
HMS *Jamaica* (C44)	Capt. Jocelyn S. C. Salter, DSO, OBE RN
90.6.2 Fire Support Unit 2	Capt. Halle C. Allan Jr., USN
USS *Mansfield* (DD-728)	Comdr. Edwin H. Headland, USN
USS *De Haven* (DD-727)	Comdr. Oscar B. Lungren, USN
USS *Lyman K. Swenson* (DD-729)	Comdr. Robert A. Schelling, USN
90.6.3 Fire Support Unit 3	Comdr. Robert H. Close, USN
USS *Collett* (DD-730)	Comdr. Robert H. Close, USN
USS *Gurke* (DD-783)	Comdr. Frederick M. Radel, USN
USS *Henderson* (DD-785)	Comdr. William S. Stewart, USN

Task Group 90.7 (Screening and Protective Group):
Capt. Richard T. Spofford, USN

Destroyers and Frigates	
USS *Rowan* (DD-782)	Comdr. Alan R. Josephson, USN
USS *Southerland* (DDR-743)	Comdr. Homer E. Conrad, USN
USS *Bayonne* (PF-21)	Lt. Comdr. Harry A. Clark, USN
USS *Newport* (PF-27)	Lt. Comdr. Percy A. Lilly Jr., USN
USS *Evansville* (PF-70)	Lt. Comdr. Eliot V. Converse Jr., USN
HMS *Mounts Bay* (F627)	Capt. John Henry Unwin, DSC RN
HMS *Whitesand Bay* (F633)	Lt. Comdr. J. V. Brothers, RN
HMNZS *Tutira* (F517)	Lt. Comdr. Peter J. H. Hoare, RN
HMNZS *Pukaki* (F424)	Lt. Comdr. Laurance E. Herrick, DSC RN
FS *La Grandiere* (F731)	Comdr. Urbain E. Cabanie, FN
Minesweepers	
USS *Pledge* (AM-277)	Lt. Richard Young, USN
USS *Partridge* (AMS-31)	Lt. (jg) Robert C. Fuller Jr., USN
USS *Mockingbird* (AMS-27)	Lt. (jg) Stanley P. Gary, USN
USS *Kite* (AMS-22)	Lt. (jg) Nicholas Grkovic, USN
USS *Osprey* (AMS-28)	Lt. (jg) Philip Levin, USN
USS *Redhead* (AMS-34)	Lt. (jg) T. R. Howard, USN
USS *Chatterer* (AMS-40)	Lt. (jg) James P. McMahon, USN

Task Force 91 (Blockade and Covering Force):
Rear Adm. William G. Andrewes, RN

HMS *Triumph* (R16)	Capt. Arthur David Torlesse, DSO RN
HMS *Ceylon* (C30)	Capt. Cromwell F. J. L. Davies, DSC RN
HMS *Cockade* (D34)	Lt. Comdr. Herbert Jack Lee, DSC RN
HMS *Charity* (D29)	Lt. Comdr. Peter R. G. Worth, DSC RN
HMCS *Cayuga* (DDE218)	Capt. Jeffery Vanstone Brock, DSC RCN
HMCS *Sioux* (DDE225)	Comdr. Paul Dalrymple Taylor, RCN
HMCS *Athabaskan* (DDE219)	Comdr. Robert Phillip Welland, DSC RCN
HMAS *Bataan* (D191)	Comdr. William B. M. Marks, RAN
HMAS *Warramunga* (D123)	Comdr. Otto H. Becher, DSC RAN
HNMS *Evertsen* (D802)	Lt. Comdr. D. J. Van Doorninck

ROK Naval Forces

Paik Doo San (PC-701)	Comdr. Chai Yong Nam, ROKN
Kum Kang San (PC-702)	Comdr. Lee Hi Jong, ROKN
Chiri San (PC-704)	Lt. Comdr. Hyun Sibak, ROKN
Tongyeong (JMS-302)	
Daegu (JMS-303)	
Danyang (JMS-306)	
Dancheon (JMS-307)	
Kang Jim (YMS-501)	
Kyong Chu (YMS-502)	
Kwang Chu (YMS-503)	
Ganggyeong (YMS-510)	
Guwolsan (YMS-512)	
Ko Yung (YMS-515)	
Yong Kung (YMS-518)[3]	

Bibliography/Notes

Boose Jr., Donald W. *Over the Beach U.S. Army Amphibious Operations in the Korean War.* Fort Leavenworth, KS: Combat Studies Institute Press U.S. Army Combined Arms Center, 2008.

Bruhn, David D. *Wooden Ships and Iron Men: The U.S. Navy's Coastal and Motor Minesweepers, 1941-1953.* Westminster, MD: Heritage Books, 2009.

Cagle, Malcolm W., Frank Albert Manson. *The Sea Services in the Korean War 1950-1953.* Annapolis, MD: Naval Institute Press and Sonalysts, Inc., 2000.

—*The Sea War in Korea.* Annapolis, MD: U.S. Naval Institute, 1957.

Edwards, Paul M. *Small United States and United Nations Warships in the Korean War.* Jefferson, NC: McFarland & Company, 2008.

Endicott, Judy G., editor. *The USAF in Korea Campaigns, Units, and Stations 1950-1953.* Maxwell Air Force Base, AL: Air Force Historical Research Agency, 2001.

Field Jr., James A. *History of United States Naval Operations: Korea.* Washington, DC: U.S. Government Printing Office, 1962.

Field, Jr., James A., Malcom Cagle, Frank A. Manson, Lynn Montross, Nicholas A. Canzona, Scott Price, Allan R. Millet. *The Sea Services in the Korea War 1950-1953.* Annapolis, MD: Naval Institute, 2000.

Haas, Michael E. *Apollo's Warriors U.S. Air Force Special Operations during the Cold War.* Maxwell Air Force Base, Alabama: Air University Press, 1997.

Inseung Kim. *The British Commonwealth and Allied Naval Forces' Operation with the Anti-Communist Guerrillas in the Korean War: With Special Reference to the Operation on the West Coast.* Dissertation submitted to the University of Birmingham for the degree of Doctor of Philosophy. University of Birmingham, May 2018.

Kyŏng-sik, Ch'oe. *The Eternal Partnership: Thailand and Korea: A History of the Participation of the Thai Forces in the Korean War.* Seoul, ROK: Ministry of Patriots & Veterans Affairs, 2010.

Lee-Hale, Peter Derek Hodgson, and Alan Ausden, *The Life and Times of H.M.S. Concord.* H.M.S. *Concord* Association, 2012.

Lott, Arnold S. *Most Dangerous Sea.* Annapolis, MD: Naval Institute, 1959.

Meid, Pat, James M. Yingling. *U.S. Marine Operations in Korea 1950-1953 Vol. V. Operations in West Korea.* Washington, DC: Historical Division Headquarters U.S. Marine Corps, 1972.

Montross, Lynn, Nicholas A. Canzona, *U. S. Marine Operations in Korea 1950-1953 Vol. II The Inchon-Seoul Operation*. Washington, DC: Historical Branch, G-3 Headquarters U. S. Marine Corps, 1955.

Perkins, J. W. *Battle Stars and Naval Awards*. Seminole, FL: Self-published, 2004.

Rottman, Gordon L. *Korean War Order of Battle: United States, United Nations, and Communist Ground, Naval, and Air Forces, 1950-1953*. Westport, CT: Praeger Publishers, 2002.

Schnabel, James F., Robert J. Watson, *History of the Joint Chiefs of Staff – The Joint Chiefs of Staff and National Policy Volume III 1950-1951 - The Korean War Part One*. Washington, DC: Office of Joint History Office of the Chairman of the Joint Chiefs of Staff, 1998.

Taplett, Robert D. *Dark Horse Six*. Williamstown, NJ: Phillips Publications, 2002.

Thorgrimsson, Thor, E. C. Russell. *Canadian Naval Operations in Korean Waters 1950-1955*. Ottawa: The Naval Historical Section Canadian Forces Headquarters Department of National Defence, 1965.

FOREWORD NOTES:

[1] "The Korean War…Defending the Friendly Islands" by Fred R. Fowlow, Starshell, Autumn 2011(https://www.navalassoc.ca/wp-content/uploads/2019/08/Starshell-Autumn-2011-Color-Final.pdf: accessed 7 October 2020).

[2] "The Korean War…Defending the Friendly Islands" by Fred R. Fowlow.

[3] Thor Thorgrimsson, E. C. Russell, *Canadian Naval Operations in Korean Waters 1950-1955* (Ottawa: The Naval Historical Section Canadian Forces Headquarters Department of National Defence, 1965), 142-143.

PREFACE NOTES:

[1] Cagle and Manson, *The Sea Services in the Korean War 1950-1953*, (Annapolis, MD: Naval Institute Press and Sonalysts, Inc., 2000), Chapter 9.

[2] Ibid.

[3] Ibid, Ch 9, Task Force 95 Established.

[4] David D. Bruhn, *Wooden Ships and Iron Men: The U.S. Navy's Coastal and Motor Minesweepers, 1941-1953* (Westminster, MD: Heritage Books, 2009), 127-128.

[5] "United Nations Forces in the Korean War" (https://anzacportal.dva.gov.au/wars-and-missions/korean-war-1950-1953/korean-war/armed-forces-korean-war/united-nations-forces-korean-war: accessed 16 August 2020).

[6] Bruhn, *Wooden Ships and Iron Men: The U.S. Navy's Coastal and Motor Minesweepers, 1941-1953*, 128-129.

[7] Ibid, 129-130.
[8] Robert D. Taplett, *Dark Horse Six* (Williamstown, NJ: Phillips Publications, 2002), 19.
[9] "U. S. Navy Special Operations in the Korean War" (https://www.history.navy.mil/research/library/online-reading-room/title-list-alphabetically/u/usnavy-special-operations-korean-war.html: accessed 17 August 2020).
[10] Ibid.
[11] "U.S. Pacific Fleet Operations - Commander in Chief U.S. Pacific Fleet - Interim Evaluation Report No. 1 - Period 25 June to 15 November 1950" (https://www.history.navy.mil/content/history/nhhc/research/library/online-reading-room/title-list-alphabetically/k/Korean-War-Interim-Evaluation-No1.html: accessed 16 August 2020).
[12] Ibid.
[13] Ibid.
[14] Ibid.
[15] "Battle Honours of RN ships & Naval Air Squadrons" (http://www.royalnavyresearcharchive.org.uk/ESCORT/Battle_hons.htm: accessed 17 August 2020).
[16] Navy and Marine Corps Awards Manual, 1953 (https://www.ibiblio.org/hyperwar/USN/ref/Awards/Awards-IV-17.html#sec2-20: accessed 17 August 2020).
[17] Ibid.
[18] J. W. Perkins, *Battle Stars and Naval Awards* (Seminole, FL: Self-published, 2004).
[19] "Patrol Frigates (PF)" (http://shipbuildinghistory.com/smallships/pf.htm: accessed 16 August 2020).
[20] Ibid.
[21] Ibid.
[22] "Patrol Craft Escort (PCE) Patrol Craft Escort (Rescue) (PCE[R]) Index" (http://www.navsource.org/archives/12/02idx.htm: accessed 15 August 2020).
[23] Ibid.
[24] Ibid.
[24] "U. S. Navy Special Operations in the Korean War" (https://www.history.navy.mil/research/library/online-reading-room/title-list-alphabetically/u/usnavy-special-operations-korean-war.html: accessed 18 August 2020).
[26] Ibid.
[27] Ibid.
[28] Ibid.
[29] Ibid.
[30] Ibid.
[31] Ibid.
[32] *Thompson, DANFS*.

[33] Thompson, *Carmick*, *DANFS*; "The Hungnam and Chinnampo Evacuations" by Edward J. Marolda (https://www.history.navy.mil/research/library/online-reading-room/title-list-alphabetically/h/the-hungnam-and-chinnampo-evacuations.html: accessed 18 August 2020).
[34] Thompson, *DANFS*; "The Hungnam and Chinnampo Evacuations" by Edward J. Marolda.
[35] "U.S. Pacific Fleet Operations - Commander in Chief U.S. Pacific Fleet - Interim Evaluation Report No. 1 - Period 25 June to 15 November 1950."
[36] Ibid.
[37] "Ships Sunk and Damaged in Action during the Korean Conflict (Partial)" (https://www.history.navy.mil/research/library/online-reading-room/title-list-alphabetically/s/ships-sunk-and-damaged-in-action-during-the-korean-conflict.html: accessed 18 August 2020).
[38] Cagle and Manson, *The Sea Services in the Korean War, 1950-1953*, Chapter 9.
[39] Pat Meid, James M. Yingling, *U.S. Marine Operations in Korea 1950-1953 Vol. V. Operations in West Korea*, 53; "The Army's Guerilla Command in Korea Part II: The Rest of the Story" by Michael E. Krivdo (https://arsofhistory.org/articles/v9n1_guerilla_comm_page_1.html: accessed 19 August 2020).
[40] "The Army's Guerilla Command in Korea Part II: The Rest of the Story" by Michael E. Krivdo
[41] "The Army's Guerilla Command in Korea Part II: The Rest of the Story" by Michael E. Krivdo; "Korean War: Chronology of U.S. Pacific Fleet Operations, January–April 1952" (https://www.history.navy.mil/content/history/nhhc/research/library/online-reading-room/title-list-alphabetically/k/korean-war-chronology/january-april-1952.html: accessed 26 March 2020).
[42] Commanding Officer and Commander Task Element 95.11, Action Report 12 July through 23 July 1952.
[43] Peter Lee-Hale, Derek Hodgson, and Alan Ausden, The Life and Times of H.M.S. Concord (H.M.S. Concord Association, 2012) (https://www.maritimequest.com/warship_directory/great_britain/photos/destroyers/concord_r63/life_and_times_of_hms_concord.pdf: accessed 20 August 2020).
[44] Guest editorial by Commander Daniel "Sandy" Herlihy, RNZN Retired, *The White Ensign Winter* 2010 (https://navymuseum.co.nz/wp-content/uploads/2012/01/white-ensign-issue-10.pdf: accessed 21 August 2020).

CHAPTER 1 NOTES:
[1] "Great Battle of Korea Strait" by Kwon Yule-jung (https://www.koreatimes.co.kr/www/opinion/2020/07/162_292097.html); "H-050-1: Naval Action in the Korean War, 25 June–1 September 1950" (https://www.history.navy.mil/content/history/nhhc/about-

us/leadership/director/directors-corner/h-grams/h-gram-050/h-050-1.html): accessed 21 July 2020.
[2] "PC-823" (http://www.navsource.org/archives/12/010823.htm: accessed 21 July 2020).
[3] Ibid.
[4] Ibid
[5] Ibid.
[6] "Great Battle of Korea Strait"; "H-050-1: Naval Action in the Korean War, 25 June–1 September 1950."
[7] "PC-823"
[8] Ibid.
[9] Federal Register Vol II Number 84, April 30, 1946; "N3-S-A1 Type" (https://cms7.marad.dot.gov/content/n3-s-a1-type); "Type N3 ship" (https://www.wikiwand.com/en/Type_N3_ship): accessed 24 July 2020.
[10] Federal Register Vol II Number 84, April 30, 1946.
[11] "Type N3 ship"; "*Alchiba* (AK-261)" (http://www.navsource.org/archives/09/13/130261.htm: accessed 24 July 2020).
[12] The Department of State bulletin Vol. XXI, No. 535, October 24, 1949.
[13] "Korea's Battleship Potemkin: Mutiny on the SS *Kimball R. Smith*" (https://koryogroup.com/blog/mutiny-on-the-ss-kimball-r-smith: accessed 23 July 2020). "Red Koreans Hold Americans To Force U.S. Recognition," *Brownsville Herald*, 16 November 1949.
[14] Ut supra.
[15] "Merchant Ships Used in the Korean War" (http://www.usmm.org/koreaships.html: accessed 23 July 2020).
[16] Robert D. Taplett, *Dark Horse Six*, 19.
[17] Bruhn, *Wooden Ships and Iron Men: The U.S. Navy's Coastal and Motor Minesweepers, 1941-1953*, 127.
[18] Ibid, 128.
[19] Ibid.
[20] Ibid, 128-129.
[21] Ibid, 129-130.
[22] Ibid, 130.
[23] "Great Battle of Korea Strait" by Kwon Yule-jung.

CHAPTER 2 NOTES:
[1] "Republic of Korea Navy" by Edward J. Marolda (https://www.history.navy.mil/research/library/online-reading-room/title-list-alphabetically/r/republic-of-korea-navy.html: accessed 26 July 2020).
[2] Gordon L. Rottman, *Korean War Order of Battle: United States, United Nations, and Communist Ground, Naval, and Air Forces, 1950-1953* (Westport, CT: Praeger Publishers, 2002), 158; "The Emerging Republic of Korea Navy – A Japanese Perspective" by Yoji Koda, *Naval War College Review*, Volume 63, Number 2, Spring 2010.

[3] "The Coast Guard and the Korean War" (https://coastguard.dodlive.mil/2010/06/the-coast-guard-and-the-korean-war/; accessed 26 July 2020).
[4] "The Emerging Republic of Korea Navy – A Japanese Perspective."
[5] "A History of the Philippine Navy in the Korean War (1950-1953)" by Mark R. Condeno (http://cimsec.org/a-history-of-the-philippine-navy-in-the-korean-war-1950-1953/37143: accessed 25 July 2020).
[6] Ibid.
[7] Ibid.
[8] "Republic of Korea Navy" by Edward J. Marolda.
[9] "Republic of Korea Navy" by Edward J. Marolda; "70 Years Ago: The 'Forgotten War' and the Fight to Hold Pusan" by Samuel Cox (https://www.maritime-executive.com/editorials/70-years-ago-the-forgotten-war-and-the-fight-to-hold-pusan: accessed 26 July 2020).
[10] "Republic of Korea Navy" by Edward J. Marolda.
[11] Bruhn, *Wooden Ships and Iron Men: The U.S. Navy's Coastal and Motor Minesweepers, 1941-1953*, 49, 54.
[12] "HMS *BYMS-2258*" (http://www.navsource.org/archives/11/19258.htm: accessed 26 July 2020).
[13] "Republic of Korea Navy" by Edward J. Marolda.
[14] Ibid.

CHAPTER 3 NOTES:

[1] James A. Field Jr., *History of United States Naval Operations: Korea*, Chapter 3 (https://www.history.navy.mil/research/library/online-reading-room/title-list-alphabetically/h/history-us-naval-operations-korea/chapter3-war-begins.html: accessed 2 March 2020).
[2] Ibid.
[3] Ibid.
[4] "HMAS *Shoalhaven*" (https://www.navy.gov.au/hmas-shoalhaven: accessed 27 July 2020).
[5] Ibid.
[6] Ibid.
[7] Ibid.
[8] Ibid.
[9] Ibid.
[10] Ibid.
[11] Ibid.
[12] Ibid.
[13] Thorgrimsson and Russell, *Canadian Naval Operations in Korean Waters 1950-1955*, 3-4.

CHAPTER 5 NOTES:

[1] Thorgimsson and Russell, *Canadian Naval Operations in Korean Waters 1950-1955*, 7.
[2] Ibid.

[3] Ibid.
[4] "H-050-1: Naval Action in the Korean War, 25 June–1 September 1950" (https://www.history.navy.mil/content/history/nhhc/about-us/leadership/director/directors-corner/h-grams/h-gram-050/h-050-1.html: accessed 5 August 2020); James F. Schnabel and Robert J. Watson, *History of the Joint Chiefs of Staff - The Joint Chiefs of Staff and National Policy Volume III 1950-1951 - The Korean War Part One* (Washington, DC: Office of Joint History Office of the Chairman of the Joint Chiefs of Staff, 1998), 37.
[5] "H-050-1: Naval Action in the Korean War, 25 June–1 September 1950"; USS *Dehaven* Sailors Association (http://ussdehaven.org/: accessed 7 August 2020).
[6] "H-050-1: Naval Action in the Korean War, 25 June–1 September 1950"; Schnabel and Watson, *History of the Joint Chiefs of Staff - The Joint Chiefs of Staff and National Policy Volume III 1950-1951 - The Korean War Part One*, 37-38.
[7] "H-050-1: Naval Action in the Korean War, 25 June–1 September 1950."
[8] James A. Field Jr., *History of United States Naval Operations: Korea*, Chapter 3.
[9] "H-050-1: Naval Action in the Korean War, 25 June–1 September 1950."
[10] "Korean War: Chronology of U.S. Pacific Fleet Operations, June-December 1950" (https://www.history.navy.mil/research/library/online-reading-room/title-list-alphabetically/k/korean-war-chronology/june-dec-1950.html: accessed 7 August 2020).
[11] Thorgimsson and Russell, *Canadian Naval Operations in Korean Waters 1950-1955*, 11.
[12] Ibid.
[13] Ibid.
[14] Inseung Kim, *The British Commonwealth and Allied Naval Forces' Operation with the Anti-Communist Guerillas in the Korean War: with Special Reference to the Operation on the West Coast*, a dissertation submitted to The University of Birmingham, May 2018 (https://etheses.bham.ac.uk/id/eprint/8169/1/Kim_Inseung18PhD.pdf: accessed 6 August 2020).
[15] Ibid.
[16] Ibid.
[17] "HMS *Cossack*" (https://www.hmscossack.co.uk/portfolio/diary/: accessed 6 August 2020).
[18] Ibid.
[19] Ibid.
[20] Ibid.
[21] "HMS *Cossack*"; Inseung Kim dissertation.
[22] "HMAS *Shoalhaven*" (https://www.navy.gov.au/hmas-shoalhaven: accessed 7 August 2020).
[23] "1949-1952: My Time Onboard the USS Collett" by Ed Shumer (http://www.usscollett.com/history/e_shumer/e_shumer%20my%20time%20aboard.htm: accessed 7 August 2020).
[24] "H-050-1: Naval Action in the Korean War, 25 June–1 September 1950."
[25] Ibid.

[26] Ibid.
[27] "French Navy in Korean Waters" by Léon C. Rochotte (https://www.koreanwar.org/html/units/un/french_navy.htm: accessed 7 August 2020).
[28] "H-050-1: Naval Action in the Korean War, 25 June–1 September 1950"; "HMAS Bataan" (https://www.navy.gov.au/hmas-bataan: accessed 9 August 2020).
[29] Thorgimsson and Russell, *Canadian Naval Operations in Korean Waters 1950-1955*, 4.
[30] Ibid, 3-4.
[31] Ibid, 11.
[32] Ibid, 12.
[33] Ibid.
[34] Ibid, 12-13.
[35] Ibid, 13.
[36] Ibid.
[37] Ibid.
[38] Ibid.
[39] "Korean War" (https://nzhistory.govt.nz/war/korean-war: accessed 7 August 2020).
[40] "Korean War" (https://erikscollectables.files.wordpress.com/2017/06/the-korean-war.pdf: accessed 29 July 2020).
[41] Ibid.
[42] James A. Field Jr., *History of United States Naval Operations: Korea*, Chapter 6 (https://www.marines.mil/Portals/1/Publications/The%20Sea%20Services%20in%20the%20Korean%20War%201950-1953%20PCN%2019000412100_1.pdf?ver=2012-10-11-164123-027: accessed 9 August 2020).
[43] Field Jr., *History of United States Naval Operations: Korea*, Chapter 7; "1945-2008 - Casualty Lists of the Royal Navy" compiled by Don Kindell (https://www.naval-history.net/xDKCas1950.htm: accessed 9 August 2020).
[44] Field Jr., *History of United States Naval Operations: Korea*, Chapter 6.
[45] Ibid.

CHAPTER 6 NOTES:
[1] Field Jr., *History of United States Naval Operations: Korea*, Chapter 7.
[2] Ibid.
[3] Bruhn, *Wooden Ships and Iron Men: The U.S. Navy's Coastal and Motor Minesweepers, 1941-1953*, 138.
[4] Ibid, 138-139.
[5] Ibid, 139.
[6] Field Jr., *History of United States Naval Operations: Korea*, Chapter 7.
[7] Ibid.
[8] Ibid.

[9] Arnold S. Lott, *Most Dangerous Sea* (Annapolis, MD: Naval Institute, 1959), 271.
[10] Ibid.
[11] Bruhn, *Wooden Ships and Iron Men: The U.S. Navy's Coastal and Motor Minesweepers, 1941-1953*, 139.
[12] Ibid.
[13] Lynn Montross and Nicholas A. Canzona, *U. S. Marine Operations in Korea 1950-1953 Vol. II The Inchon-Seoul Operation* (Washington, DC: Historical Branch, G-3 Headquarters U. S. Marine Corps, 1955), Appendix E: Task Organization Joint Task Force Seven
[14] Bruhn, *Wooden Ships and Iron Men: The U.S. Navy's Coastal and Motor Minesweepers, 1941-1953*, 139.
[15] Field Jr., *History of United States Naval Operations: Korea*, Chapter 7.
[16] Ibid.
[17] "History of the USS *Lyman K. Swenson* DD729" (http://www.dd729.com/hist.html: accessed 9 August 2020).
[18] Field Jr., *History of United States Naval Operations: Korea*, Chapter 7; "History of the USS *Lyman K. Swenson* DD729."
[19] Ut supra.
[20] Montross and Canzona, *U. S. Marine Operations in Korea 1950-1953 Vol. II The Inchon-Seoul Operation*.
[21] "Mine Warfare Korea" by Léon C. Rochotte (http://kman.my.meganet.net/nfleon.htm: accessed 9 August 2020).
[22] Thorgimsson and Russell, *Canadian Naval Operations in Korean Waters 1950-1955*, 17.
[23] Montross and Canzona, *U. S. Marine Operations in Korea 1950-1953 Vol. II The Inchon-Seoul Operation*, Appendix E: Task Organization Joint Task Force Seven.
[24] Thorgimsson and Russell, *Canadian Naval Operations in Korean Waters 1950-1955*, 17.
[25] Ibid.
[26] "HMAS *Warramunga* (I)" (https://www.navy.gov.au/hmas-warramunga-i: accessed 10 September 2020).
[27] Lott, *Most Dangerous Sea*, 271.
[28] Ibid, 268.
[29] Bruhn, *Wooden Ships and Iron Men: The U.S. Navy's Coastal and Motor Minesweepers, 1941-1953*, 139-140.
[30] Field Jr., *History of United States Naval Operations: Korea*, Chapter 7.
[31] Ibid.
[32] Montross and Canzona, *U. S. Marine Operations in Korea 1950-1953 Vol. II The Inchon-Seoul Operation*, 133, 321.
[33] General Orders: Commander 7th Fleet: Serial 1090 (November 20, 1950); "David H. Swenson, Jr., LTJG, USN" (https://usnamemorialhall.org/index.php/DAVID_H._SWENSON,_JR.,_LTJG,_USN: accessed 10 August 2020).

CHAPTER 7 NOTES:
[1] Inseung Kim dissertation.
[2] Ibid.
[3] Ibid.
[4] Ibid.
[5] Thorgrimsson and Russell, *Canadian Naval Operations in Korean Waters 1950-1955*, 18.
[6] Ibid.
[7] Ibid.
[8] Ibid.
[9] Ibid, 19.
[10] Ibid.
[11] Ibid.
[12] "Awards to the Royal Canadian Navy for Korea" (https://www.blatherwick.net/documents/Korean%20War%20Honours%20to%20Canadians/RCN%20awards%20for%20Korea.pdf: accessed 23 August 2020).
[12] Thorgrimsson and Russell, *Canadian Naval Operations in Korean Waters 1950-1955*, 19
[13] Ibid.
[14] Thorgrimsson and Russell, *Canadian Naval Operations in Korean Waters 1950-1955*, 19; Inseung Kim dissertation.
[15] Inseung Kim dissertation.
[16] Ibid.
[17] "A chronology of the Korean War" (https://anzacportal.dva.gov.au/wars-and-missions/korean-war-1950-1953/korean-war/cold-war-and-crisis-korea/chronology-korean-war): accessed 23 August 2020.
[18] "Chronology of U.S. Pacific Fleet Operations, June-December 1950" (https://www.history.navy.mil/research/library/online-reading-room/title-list-alphabetically/k/korean-war-chronology.html: assessed 23 August 2020).
[19] Inseung Kim dissertation; "Historical Calendar – 1950" (https://www.veterans.gc.ca/eng/remembrance/history/korean-war/land-morning-calm/events/1950: accessed 23 August 2020).
[20] Inseung Kim dissertation.
[21] Ibid.
[22] Ibid.
[23] Ibid.
[24] Bruhn, *Wooden Ships and Iron Men: The U.S. Navy's Coastal and Motor Minesweepers, 1941-1953*, 142-143.
[25] Ibid.
[26] Ibid.
[27] "Mine Warfare" by Edward J. Marolda (https://www.history.navy.mil/content/history/nhhc/research/library/online-reading-room/title-list-alphabetically/m/mine-warfare.html: accessed 24 August 2020).

[28] Bruhn, *Wooden Ships and Iron Men: The U.S. Navy's Coastal and Motor Minesweepers, 1941-1953*, 158, 160.
[29] Ibid.
[30] "Stephen Morris Archer 18 July 1911-31 July 1996" (https://www.history.navy.mil/research/library/research-guides/modern-biographical-files-ndl/modern-bios-a/archer-stephen-morris.html: accessed 25 August 2020).
[31] General Orders: Commander 7th Fleet: Serial 1204 (December 13, 1950).
[32] "Mine Warfare" by Edward J. Marolda.
[33] Ibid.
[34] Ibid.
[35] "Sikorsky Helicopters Came of Age in the Korean War" (https://www.sikorskyarchives.com/pdf/news_2017/News_Oct_2017.pdf: accessed 24 August 2020).
[36] Lott, *Most Dangerous Sea*, 279.
[37] Ibid.
[38] Ibid, 279-280.
[39] "Stephen Morris Archer 18 July 1911-31 July 1996."
[40] *Horace A. Bass, DANFS*; "CIA Paramilitary Operations Korea, 1950 - 1951" by Charles H. Briscoe (https://arsof-history.org/articles/pdf/v9n1_cia_paramilitary.pdf: accessed 25 August 2020).

CHAPTER 8 NOTES:
[1] Donald W. Boose Jr., *Over the Beach U.S. Army Amphibious Operations in the Korean War* (Fort Leavenworth, KS: Combat Studies Institute Press U.S. Army Combined Arms Center, 2008) 234 (https://www.armyupress.army.mil/Portals/7/combat-studies-institute/csi-books/boose.pdf: accessed 28 August 2020).
[2] Ibid.
[3] Boose Jr., *Over the Beach U.S. Army Amphibious Operations in the Korean War*, 234; Field Jr., *History of United States Naval Operations: Korea*, Chapter 9.
[4] Ibid.
[5] Ibid.
[6] Ibid.
[7] Boose Jr., *Over the Beach U.S. Army Amphibious Operations in the Korean War*, 238; Field Jr., *History of United States Naval Operations: Korea*, Chapter 9.
[8] Field Jr., *History of United States Naval Operations: Korea*, Chapter 9.
[9] Ibid.
[10] Thorgrimsson and Russell, *Canadian Naval Operations in Korean Waters 1950-1955*, 29.
[11] Ibid, 31.
[12] Ibid.
[13] Ibid.
[14] Ibid.
[15] Ibid.

[16] Ibid, 32.
[17] Ibid.
[18] Ibid.
[19] Thorgrimsson and Russell, *Canadian Naval Operations in Korean Waters 1950-1955*, 32; "Awards to the Royal Canadian Navy for Korea."
[20] Thorgrimsson and Russell, *Canadian Naval Operations in Korean Waters 1950-1955*, 33.
[21] Ibid.
[22] Ibid.
[23] Ibid, 34.
[24] Ibid.
[25] Ibid.
[26] Ibid.
[27] Ibid, 34-35.
[28] Thorgrimsson and Russell, *Canadian Naval Operations in Korean Waters 1950-1955*, 35; "United Nations and Republic of Korea Forces -- Japanese Contributions" (https://www.ibiblio.org/hyperwar/OnlineLibrary/photos/events/kowar/un-rok/jpn.htm: accessed 6 December 2020).
[29] Ibid.
[30] "Awards to the Royal Canadian Navy for Korea."
[31] Military Times Hall of Valor (https://valor.militarytimes.com/: accessed 29 August 2020).

CHAPTER 9 NOTES:
[1] "Mine Warfare Korea" by Léon C. Rochotte (http://kman.my.meganet.net/nfleon.htm: accessed 29 August 2020).
[2] "Mine Warfare Korea" by Léon C. Rochotte; Lott, *Most Dangerous Sea*, 272.
[3] "Mine Warfare Korea" by Léon C. Rochotte.
[4] Ibid.
[5] Ibid.
[6] "Cao Bang, Battle of" (http://indochine.uqam.ca/en/historical-dictionary/206-cao-bng-battle-of.html: accessed 30 August 2020).
[7] Ibid.
[8] Ibid.
[9] Ibid.
[10] Ch'oe.Kyŏng-sik, *The Eternal Partnership: Thailand and Korea: A History of the Participation of the Thai Forces in the Korean War* (Seoul, ROK: Ministry of Patriots & Veterans Affairs, 2010), 108 (https://web.archive.org/web/20140826120037/http://english.mpva.go.kr/upload/Contents/2010thailand.pdf: accessed 13 August 2020).
[11] "Thailand" (https://www.mpva.go.kr/english/front/cooperation/unallies_Thailand02.do: accessed 13 August 2020).

CHAPTER 10 NOTES:
[1] Inseung Kim dissertation.
[2] Ibid.
[3] Ibid.
[4] Ibid.
[5] Ibid.
[6] Ibid.
[7] Boose Jr., *Over the Beach U.S. Army Amphibious Operations in the Korean War*, 256-257.
[8] Ibid.
[9] Ibid, 257.
[10] Inseung Kim dissertation.
[11] Ibid.
[12] Ibid.
[13] Ibid.
[14] Ibid.
[15] Field Jr., *History of United States Naval Operations: Korea*, Chapter 9.
[16] Ibid.
[17] Ibid.
[18] Ibid.
[19] Ibid.
[20] Ibid.
[21] Ibid.
[22] Ibid.
[23] Ibid.
[24] Ibid.
[25] Ibid.
[26] Inseung Kim dissertation.
[27] Ibid.
[28] Ibid.
[29] Boose Jr., *Over the Beach U.S. Army Amphibious Operations in the Korean War*, 257.
[30] Ibid, 257-258.
[31] Ibid, 259.
[32] Ibid.
[33] Ibid.

CHAPTER 11 NOTES:
[1] Inseung Kim dissertation.
[2] Thorgimsson and Russell, *Canadian Naval Operations in Korean Waters 1950-1955*, 40.
[3] Ibid, 41.
[4] "Korean War: Chronology of U.S. Pacific Fleet Operations, January–June 1951" (https://www.history.navy.mil/research/library/online-reading-room/title-list-alphabetically/k/korean-war-chronology/january-june-1951.html: accessed 3 September 2020).

[5] Thorgimsson and Russell, *Canadian Naval Operations in Korean Waters 1950-1955*, 41-42.
[6] Ibid, 42.
[7] Cagle and Manson, *The Sea Services in the Korean War 1950-1953*, Chapter 9.
[8] Cagle and Manson, *The Sea Services in the Korean War 1950-1953*, Chapter 9; "Korean War: Chronology of U.S. Pacific Fleet Operations, January–June 1951."
[9] "Korean War: Chronology of U.S. Pacific Fleet Operations, January–June 1951."
[10] Inseung Kim dissertation.
[11] Ibid.
[12] Ibid.
[13] Ibid.
[14] Ibid.
[15] Ibid.
[16] Ibid.
[17] Ibid.
[18] Ibid.
[19] Ibid.
[20] Boose Jr., *Over the Beach U.S. Army Amphibious Operations in the Korean War*, 267.
[21] Ibid.
[22] Ibid, 267-268.
[23] Ibid.
[24] Ibid.
[25] Ibid.
[26] Ibid.
[27] Ibid.
[28] Ibid.
[29] "HMAS *Warramunga* (I)" (https://www.navy.gov.au/hmas-warramunga-i: accessed 4 September 2020).
[30] Ibid.
[31] Inseung Kim dissertation.
[32] Ibid.
[33] Ibid.
[34] Boose Jr., *Over the Beach U.S. Army Amphibious Operations in the Korean War*, 294.
[35] "A Combat First Army SF Soldiers in Korea, 1953-1955" by Kenneth Finlayson (https://arsof-history.org/articles/v9n1_sf_in_korea_page_1.html: accessed 21 August 2020).
[36] Boose Jr., *Over the Beach U.S. Army Amphibious Operations in the Korean War*, 294.

CHAPTER 12 NOTES:

[1] Cagle and Manson, *The Sea Services in the Korean War 1950-1953*, Chapter 9.
[2] Ibid.

[3] Field Jr., *History of United States Naval Operations: Korea*, Chapter 10. (https://www.history.navy.mil/content/history/nhhc/research/library/online-reading-room/title-list-alphabetically/h/history-us-naval-operations-korea/chapter10-second-six-months.html: accessed 4 September 2020).
[4] Ibid.
[5] Ibid.
[6] Ibid.
[7] Ibid.
[8] Ibid.
[9] Ibid.
[10] Ibid.
[11] Ibid.
[12] Ibid.
[13] Ibid.
[14] Commanding Officer, USS *Bataan* (CVL 29), Action Report; period 12 May 1951-13 June 1951; submission of, [day of month obscured] June 1951.
[15] Ibid.
[16] Ibid.
[17] Ibid.
[18] Ibid.
[19] Ibid.
[20] Ibid.
[21] Field Jr., *History of United States Naval Operations: Korea*, Chapter 10.

CHAPTER 13 NOTES
[1] "Korea and the RNZN" (http://navymuseum.co.nz/korea-and-the-rnzn/: accessed 9 June 2020).
[2] Ibid.
[3] Ibid.
[4] "Korea and the RNZN."
[5] Ibid.
[6] Ibid.
[7] Ibid.
[8] *"Tutira"* (https://navymuseum.co.nz/explore/by-collections/ships/tutira-loch-class-frigate/#: accessed 19 September 2020).
[9] "Bob Marchioni" by Neville Peach, *The White Ensign* Royal New Zealand Navy Museum Journal, Issue 10 Winter 2010 (https://navymuseum.co.nz/wp-content/uploads/2012/01/white-ensign-issue-10.pdf: accessed 18 September 2020).
[10] *"Pukaki"* – Frigate" (https://navymuseum.co.nz/explore/by-collections/ships/pukaki-loch-class-frigate/: accessed 19 September 2020).
[11] "Bob Marchioni" by Neville Peach, *The White Ensign* Royal New Zealand Navy Museum Journal, Issue 10 Winter 2010.
[12] "41 Independent Commando R.M. Korea 1950-1952" (http://www.koreanwaronline.com/arms/41RMCpub.htm); "41 Independent Commando RM"

(http://www.commandoveterans.org/41IndependentCommando): accessed 19 September 2020).
[13] "Bob Marchioni" by Neville Peach, *The White Ensign*, Winter 2010.
[14] "Bob Marchioni" by Neville Peach, *The White Ensign*, Winter 2010; "Missing in North Korea: Ian McGibbon explains the circumstances that led to a New Zealand serviceman becoming one of the Korean War missing near the northern port of Chinnampo"
(https://www.thefreelibrary.com/Missing+in+North+Korea%3A+Ian+McGibbon+explains+the+circumstances+that...-a0277435197: accessed 19 September 2020).
[15] Ut supra.
[16] Ut supra.
[17] "Missing in North Korea: Ian McGibbon explains the circumstances that led to a New Zealand serviceman becoming one of the Korean War missing near the northern port of Chinnampo."
[18] "Bob Marchioni" by Neville Peach, *The White Ensign*, Winter 2010; "Missing in North Korea: Ian McGibbon explains the circumstances that led to a New Zealand serviceman becoming one of the Korean War missing near the northern port of Chinnampo."
[19] "Missing in North Korea: Ian McGibbon explains the circumstances that led to a New Zealand serviceman becoming one of the Korean War missing near the northern port of Chinnampo."
[20] Ibid.
[21] Ibid.
[22] "Bob Marchioni" by Neville Peach, *The White Ensign*, Winter 2010.
[23] "Missing in North Korea: Ian McGibbon explains the circumstances that led to a New Zealand serviceman becoming one of the Korean War missing near the northern port of Chinnampo."

CHAPTER 14 NOTES:
[1] Cagle and Manson, *The Sea Services in the Korean War 1950-1953*, Chapter 9.
[2] Cagle and Manson, *The Sea Services in the Korean War 1950-1953*, Chapter 9; Field Jr., *History of United States Naval Operations: Korea*, Chapter 12.
(https://www.history.navy.mil/content/history/nhhc/research/library/online-reading-room/title-list-alphabetically/h/history-us-naval-operations-korea/chapter12-two-more-years.html: accessed 6 September 2020).
[3] "Across the Pacific to War, The Colombian Navy in Korea, 1951–1955" by Charles H. Briscoe (https://arsof-history.org/articles/v2n4_across_pacific_page_1.html: accessed 28 July 2020).
[4] Ibid.
[5] Ibid.
[6] Ibid.
[7] Ibid.
[8] Ibid.
[9] Field Jr., *History of United States Naval Operations: Korea*, Chapter 12.

[10] Ibid.
[11] Ibid.
[12] Field Jr., *History of United States Naval Operations: Korea*, Chapter 12; *Whetstone, DANFS*; "HMS *Cardigan Bay* (K 630) - Bay-class Frigate (https://www.naval-history.net/xGM-Chrono-15Fr-Bay-HMS_Cardigan_Bay.htm: accessed 6 September 2020).
[13] Ut supra.
[14] Ut supra.
[15] Field Jr., *History of United States Naval Operations: Korea*, Chapter 12.
[16] Ibid.
[17] Cagle and Manson, *The Sea Services in the Korean War 1950-1953*, Chapter 9.
[18] "HMAS *Murchison*" (https://www.navy.gov.au/hmas-murchison: accessed 13 August 2020); Cagle and Manson, *The Sea Services in the Korean War 1950-1953*, Chapter 9.
[19] Ut supra.
[20] "HMAS *Murchison*"; Cagle and Manson, *The Sea Services in the Korean War 1950-1953*, Chapter 9; "Baron Murchison of Han" by Ronald McKie (http://www.gunplot.net/main/content/hmas-murchison-baron-han: accessed 7 September 2020).
[21] "HMAS *Murchison*"; Cagle and Manson, *The Sea Services in the Korean War 1950-1953*, Chapter 9.
[22] Field Jr., *History of United States Naval Operations: Korea*, Chapter 12.
[23] "Baron Murchison of Han" by Ronald McKie.
[24] Cagle and Manson, *The Sea Services in the Korean War 1950-1953*, Chapter 9.
[25] Ibid.
[26] Field Jr., *History of United States Naval Operations: Korea*, Chapter 12.
[27] Cagle and Manson, *The Sea Services in the Korean War 1950-1953*, Chapter 9.
[28] "Dollard, Allen Nelson (Et) (1917–2009)" by Michael Fogarty (http://oa.anu.edu.au/obituary/dollard-allen-nelson-et-314: accessed 7 September 2020).

CHAPTER 15 NOTES:
[1] Field Jr., *History of United States Naval Operations: Korea*, Chapter 12.
[2] Field Jr., *History of United States Naval Operations: Korea*, Chapter 12; "Wolfpacks and Donkeys Special Forces Soldiers in the Korean War" by Kenneth Finlayson.
[3] Field Jr., *History of United States Naval Operations: Korea*, Chapter 12.
[4] Field Jr., *History of United States Naval Operations: Korea*, Chapter 12; "The Evolution of U.S. Army HUMINT: Intelligence Operations in the Korean War" by John P. Finnegan (https://www.cia.gov/library/center-for-the-study-of-intelligence/csi-publications/csi-studies/studies/vol.-55-no.-2/pdfs-vol.-55-no.-2/Finnegan-HUMINT%20in%20the%20Korean%20War-7%20June2011.pdf: accessed 8 September 2020).
[5] "The Evolution of U.S. Army HUMINT: Intelligence Operations in the Korean War" by John P. Finnegan.

[6] Ibid.
[7] Ibid.
[8] Field Jr., *History of United States Naval Operations: Korea*, Chapter 12; Michael E. Haas, *Apollo's Warriors U.S. Air Force Special Operations during the Cold War* (Maxwell Air Force Base, Alabama: Air University Press, 1997), 67-74 (https://www.airuniversity.af.edu/Portals/10/AUPress/Books/b_0037_haas_apollos_warriors.pdf: accessed 8 September 2020).
[9] Haas, *Apollo's Warriors U.S. Air Force Special Operations during the Cold War*, 67-74; Judy G. Endicott, editor, *The USAF in Korea Campaigns, Units, and Stations 1950 –1953* (Maxwell Air Force Base, AL: Air Force Historical Research Agency, 2001), 109 (https://media.defense.gov/2010/May/26/2001330297/-1/-1/0/AFD-100526-045.pdf: accessed 8 September 2020).
[10] "Air War Korea, 1950-53" by A. Timothy Warnock (https://www.airforcemag.com/article/1000korea/: accessed 8 September 2020).
[11] Ibid.
[12] Field Jr., *History of United States Naval Operations: Korea*, Chapter 12.
[13] Thorgimsson and Russell, *Canadian Naval Operations in Korean Waters 1950-1955*, 60.
[14] Ibid.
[15] Ibid, 61.
[16] Ibid.
[17] Ibid.
[18] Ibid.
[19] Ibid, 60, 63-64.
[20] Ibid, 64.
[21] Thorgimsson and Russell, *Canadian Naval Operations in Korean Waters 1950-1955*, 64; "The Case of the Spurious Sawbones" by Les Peate (http://www.kvacanada.com/stories_lpimposter.htm: accessed 10 August 2020).
[22] "Ferdinand Waldo (Fred) Demara" by Captain (N) (Ret'd) M. Braham, CD, Fact Sheet "41 published by The Friends of the Canadian War Museum.
[23] "Ferdinand Waldo (Fred) Demara" by Captain (N) (Ret'd) M. Braham; "The Case of the Spurious Sawbones" by Les Peate.
[24] "Ferdinand Waldo (Fred) Demara" by Captain (N) (Ret'd) M. Braham.
[25] "Ferdinand Waldo (Fred) Demara" by Captain (N) (Ret'd) M. Braham; "The Case of the Spurious Sawbones" by Les Peate.
[26] Ut supra.
[27] "Ferdinand Waldo (Fred) Demara" by Captain (N) (Ret'd) M. Braham.

CHAPTER 16 NOTES:
[1] Inseung Kim dissertation.
[2] Ibid.
[3] Ibid.
[4] Ibid.

[5] Ibid.
[6] Ibid.
[7] Inseung Kim dissertation; "Korean War: Chronology of U.S. Pacific Fleet Operations, July–December 1951" (https://www.history.navy.mil/research/library/online-reading-room/title-list-alphabetically/k/korean-war-chronology/july-december-1951.html: accessed 11 September 2020).
[8] Thorgrimsson and Russell, *Canadian Naval Operations in Korean Waters 1950-1955*, 71-72.
[9] Ibid, 72.
[10] Ibid.
[11] Ibid, 73.
[12] Ibid.
[13] Thorgrimsson and Russell, *Canadian Naval Operations in Korean Waters 1950-1955*, 73.
[14] "James Plomer DSC, RCNVR" (https://uboat.net/allies/commanders/2014.html); "Plomer, James" (http://www.nauticapedia.ca/dbase/Query/Biolist3.php?name=Plomer,%20James&id=7257&Page=1&input=Plomer,%20James): both accessed 12 September 2020.
[15] Thorgrimsson and Russell, *Canadian Naval Operations in Korean Waters 1950-1955*, 77.
[16] Ibid.
[17] Thorgrimsson and Russell, *Canadian Naval Operations in Korean Waters 1950-1955*, 77; "1945-2008 - Casualty Lists of the Royal Navy" compiled by Don Kindell.
[18] Thorgrimsson and Russell, *Canadian Naval Operations in Korean Waters 1950-1955*, 77.
[19] Ibid.
[20] Ibid.
[21] Ibid.
[22] Ibid.
[23] Ibid, 77-78.
[24] Ibid, 78.
[25] Ibid.
[26] Ibid, 79.
[27] Ibid.
[28] Ibid, 145.
[29] Ibid, 79.
[30] Field Jr., *History of United States Naval Operations: Korea*, Chapter 12.
[31] Ibid.
[32] Field Jr., *History of United States Naval Operations: Korea*, Chapter 12; "Korean War: Chronology of U.S. Pacific Fleet Operations, July–December 1951."
[33] Ut supra.

[34] "Korean War: Chronology of U.S. Pacific Fleet Operations, July–December 1951."

CHAPTER 17 NOTES:
[1] "HMCS *Athabaskan* 219 Shipboard Newspaper Christmas Edition - December 1951" (http://www.forposterityssake.ca/RCN-DOCS/ATH-XMAS-1951.htm: accessed 22 September 2020).
[2] "Semaphore: The Navy and the British Commonwealth Occupation Force in Japan 1945-1952" by Greg Swinden (https://www.navy.gov.au/media-room/publications/semaphore-navy-and-british-commonwealth-occupation-force-japan-1945-1952); "Korean souvenirs... aboard HMS Charity, 1952... excerpts of Edward Bates Memories" (https://france-coree.pagesperso-orange.fr/eurokorvet/uk/ted_charity.htm): both accessed 21 September 2020.
[3] "Semaphore: The Navy and the British Commonwealth Occupation Force in Japan 1945-1952" by Greg Swinden.
[4] "Logistics & Support Activities in Japan, 1950-1953 – Overview and Selected Views" (https://www.ibiblio.org/hyperwar/OnlineLibrary/photos/events/kowar/log-sup/japan.htm: accessed 22 September 2020).
[5] Ibid.
[6] "HMCS *Athabaskan* 219 Shipboard Newspaper Christmas Edition - December 1951."
[7] Ibid.
[8] "Heart of Oak – March of the Royal Canadian Navy" (https://hamiltonnaval.ca/heart-of-oak-march-of-the-royal-canadian-navy/: accessed 22 September 2020).
[9] "HMCS *Athabaskan* 219 Shipboard Newspaper Christmas Edition - December 1951"; "1947 – 1950 Camp Wood Japan" (https://gimlets4ever.wordpress.com/1945/10/17/1947-1950-camp-wood-japan/: accessed 20 September 2020).
[10] "1947 – 1950 Camp Wood Japan."
[11] "HMCS *Athabaskan* 219 Shipboard Newspaper Christmas Edition - December 1951."
[12] Ibid.
[13] Ibid.
[14] Ibid.
[15] Ibid.

CHAPTER 18 NOTES:
[1] ADM 1/23906, Korean Naval Operations-9 January to 8 February 1952, 6.
[2] Thorgimsson and Russell, *Canadian Naval Operations in Korean Waters 1950-1955*, 87; "Korean War: Chronology of U.S. Pacific Fleet Operations, January–April 1952" (https://www.history.navy.mil/research/library/online-reading-room/title-list-alphabetically/k/korean-war-chronology/january-april-1952.html); "The Korean War Hasn't Officially Ended. One Reason:

POWs" (https://www.history.com/news/korean-war-peace-treaty-pows): accessed 14 September 2020).
[3] Thorgimsson and Russell, *Canadian Naval Operations in Korean Waters 1950-1955*, 87.
[4] Ibid.
[5] Ibid, 87-88.
[6] Peter Lee-Hale, Derek Hodgson, and Alan Ausden, *The Life and Times of H.M.S. Concord*.
[7] Inseung Kim dissertation.
[8] Ibid
[9] Ibid.
[10] Ibid.
[11] Ibid.
[12] Ibid.
[13] Thorgimsson and Russell, *Canadian Naval Operations in Korean Waters 1950-1955*, 88.
[14] Ibid.
[15] Ibid.
[16] Ibid.
[17] Ibid.
[18] Ibid.
[19] Ibid.
[20] "The Life and Times of H.M.S. *Concord*."
[21] Ibid.
[22] Ibid.
[23] Ibid.
[24] "Lieutenant Augustus Willington Shelton Agar VC" (https://www.suffolkarchives.co.uk/times/war-and-conflict/lieutenant-augustus-willington-shelton-agar-vc/: accessed 28 September 2020).
[25] Ibid.
[26] Ibid.
[27] Ibid.
[28] "Lieutenant Augustus Willington Shelton Agar VC"; "Royal Navy (RN) Officers 1939-1945"
(https://www.unithistories.com/officers/RN_officersA2.html: accessed 28 September 2020).
[29] "Lieutenant Augustus Willington Shelton Agar VC."
[30] Ibid.
[31] Ibid.

CHAPTER 19 NOTES:
[1] Thorgrimsson and Russell, *Canadian Naval Operations in Korean Waters 1950-1955*, 100.
[2] Ibid, 99-100.
[3] Ibid.
[4] Ibid.

[5] Ibid.
[6] Ibid, 99-100.
[7] Ibid, 100.
[8] Ibid.
[9] Ibid.
[10] Ibid.
[11] Thorgrimsson and Russell, *Canadian Naval Operations in Korean Waters 1950-1955*, 100; Commanding Officer and Commander Task Element 95.11, Action Report 5 June through 16 June 1952, 10 July 1952.
[12] Ut supra.
[13] Thorgrimsson and Russell, *Canadian Naval Operations in Korean Waters 1950-1955*, 101.
[14] Ibid.
[15] Ibid.
[16] Ibid.
[17] Ibid.
[18] Commanding Officer and Commander Task Element 95.11, Action Report 5 June through 16 June 1952, 10 July 1952.
[19] Thorgrimsson and Russell, *Canadian Naval Operations in Korean Waters 1950-1955*, 101.
[20] "The Army's Guerilla Command in Korea Part II: The Rest of the Story" by Michael E. Krivdo (https://arsof-history.org/articles/v9n1_guerilla_comm_page_1.html: accessed 26 March 2020).
[21] Commanding Officer and Commander Task Element 95.11, Action Report 12 July through 23 July 1952.
[22] Pat Meid, James M. Yingling, *U.S. Marine Operations in Korea 1950-1953 Vol. V. Operations in West Korea*, 53; "The Army's Guerilla Command in Korea Part II: The Rest of the Story" by Michael E. Krivdo.
[23] "The Army's Guerilla Command in Korea Part II: The Rest of the Story" by Michael E. Krivdo.
[24] "Korean War: Chronology of U.S. Pacific Fleet Operations, May–August 1952" (http://overlord-wot.blogspot.com/2017_10_15_archive.html: accessed 26 July 2020).
[25] Commanding Officer and Commander Task Element 95.11, Action Report 12 July through 23 July 1952.
[26] "Wolfpacks and Donkeys Special Forces Soldiers in the Korean War" by Kenneth Finlayson.
[27] Thorgrimsson and Russell, *Canadian Naval Operations in Korean Waters 1950-1955*, 107.
[28] "Wolfpacks and Donkeys Special Forces Soldiers in the Korean War."
[29] Thorgrimsson and Russell, *Canadian Naval Operations in Korean Waters 1950-1955*, 107.
[30] John MacFarlane email of 26 September 2020.
[31] Thorgrimsson and Russell, *Canadian Naval Operations in Korean Waters 1950-1955*, 108.

[32] Commanding Officer, USS *Sicily*, Action Report for the period of 4 September through 13 September 1952, 13 November 1952.
[33] Thorgrimsson and Russell, *Canadian Naval Operations in Korean Waters 1950-1955*, 108.
[34] Ibid.
[35] Ibid.
[36] Commanding Officer, USS *Sicily*, Action Report for the period of 4 September through 13 September 1952, 13 November 1952.
[37] Thorgrimsson and Russell, *Canadian Naval Operations in Korean Waters 1950-1955*, 108.
[38] Ibid.
[39] Thorgrimsson and Russell, *Canadian Naval Operations in Korean Waters 1950-1955*, 108; Commanding Officer, USS *Sicily*, Action Report for the period of 4 September through 13 September 1952, 13 November 1952.

CHAPTER 20 NOTES:
[1] "Bogey on the Radar" by Don M. Jatiouk (http://jproc.ca/nootka/bogey.html: accessed 13 September 2020).
[2] Thorgimsson and Russell, *Canadian Naval Operations in Korean Waters 1950-1955*, 108-109.
[3] Ibid.
[4] Ibid, 109.
[5] Thorgimsson and Russell, *Canadian Naval Operations in Korean Waters 1950-1955*, 109; "Bogey on the Radar" by Don M. Jatiouk.
[6] Thorgimsson and Russell, *Canadian Naval Operations in Korean Waters 1950-1955*, 109.
[7] Ibid.
[8] Ibid, 109-110.
[9] Ibid.
[10] Ibid.
[11] "Korean War: Chronology of U.S. Pacific Fleet Operations, January–April 1952"; "The Naval Nemesis in Korea" by Charles H. Briscoe (https://arsof-history.org/articles/v9n1_cia_paramilitary_sb_nemesis.html: accessed 18 September 2020).
[12] "The Naval Nemesis in Korea" by Charles H. Briscoe.
[13] Thorgimsson and Russell, *Canadian Naval Operations in Korean Waters 1950-1955*, 104.
[14] Ibid, 106
[15] Ibid, 104-105.
[16] "Royal Navy (Rum Ration)" (https://hansard.parliament.uk/Commons/1970-01-28/debates/59960fac-6112-4c16-b660-04fef75dbf9b/RoyalNavy(RumRation): accessed 14 September 2020).
[17] Ibid.

[18] "On this day 1970…the last rum ration" by Rupert Millar (https://www.thedrinksbusiness.com/2015/07/on-this-day-1970the-last-rum-ration/: accessed 14 September 2020).
[19] "The History of the Tot" (http://www.readyayeready.com/tradition/tot-history.php: accessed 21 September 2020).
[20] Ibid.
[21] Ibid.
[22] Ibid.
[23] "Jackspeak of the Royal Canadian Navy" (http://www.readyayeready.com/jackspeak/termview.php?id=1812: accessed 21 September 2020).

CHAPTER 21 NOTES:
[1] "Korean War: Chronology of U.S. Pacific Fleet Operations, September–December 1952" (https://www.history.navy.mil/content/history/nhhc/research/library/online-reading-room/title-list-alphabetically/k/korean-war-chronology/september-december-1952.html: accessed 30 September 2020).
[2] Ibid.
[3] "HMAS *Anzac* (II)" (https://www.navy.gov.au/hmas-anzac-ii: accessed 30 September 2020); "Korean War: Chronology of U.S. Pacific Fleet Operations, September–December 1952."
[4] "Korean War: Chronology of U.S. Pacific Fleet Operations, September–December 1952."
[5] Ibid.
[6] Ibid.
[7] Ibid.
[8] Ibid.
[9] "Korean War: Chronology of U.S. Pacific Fleet Operations, January - April 1953" (https://www.history.navy.mil/research/library/online-reading-room/title-list-alphabetically/k/korean-war-chronology/january-april-1953.html: accessed 30 September 2020).
[10] Ibid.
[11] Ibid.
[12] Ibid.
[13] Ibid.
[14] "Korean War: Chronology of U.S. Pacific Fleet Operations, January - April 1953"; "Korean war" (https://erikscollectables.files.wordpress.com/2017/06/the-korean-war.pdf: accessed 12 August 2020).
[15] "Korean war."
[16] "The Long Road to the Korean War Armistice" (https://thediplomat.com/2018/08/the-long-road-to-the-korean-war-armistice/: accessed 1 October 2020).
[17] Ibid.

[18] "Korean War: Chronology of U.S. Pacific Fleet Operations, May–July 1953"
(https://www.history.navy.mil/content/history/nhhc/research/library/online-reading-room/title-list-alphabetically/k/korean-war-chronology/may-july-1953.html: accessed 1 October 2020).
[19] Ibid.
[20] Ibid.
[21] Ibid.
[22] Ibid.
[23] Ibid.
[24] Ibid.
[25] Ibid.
[26] Ibid.
[27] Ibid.

CHAPTER 22 NOTES:
[1] "The Korean War Armistice Agreement"
(https://www.usfk.mil/Portals/105/Documents/SOFA/G_Armistice_Agreement.pdf: accessed 26 September 2020).
[2] Inseung Kim dissertation.
[3] Ibid.
[4] Ibid.
[5] Ibid.
[6] Ibid.
[7] "The Far East and a New War" by William Henry Cook
(http://billcook.hostfree.pw/korea.html?i=1: accessed 26 September 2020).
[8] Paul M. Edwards, *Small United States and United Nations Warships in the Korean War* (Jefferson, NC: McFarland & Company, 2008), 212; "The Far East and a New War" by William Henry Cook
(http://billcook.hostfree.pw/korea.html?i=1); "HMS *Birmingham* (19)"
(https://uboat.net/allies/warships/ship/1228.html): accessed 26 September 2020.
[9] "Korean War: Chronology of U.S. Pacific Fleet Operations, May–July 1953."
(https://www.history.navy.mil/research/library/online-reading-room/title-list-alphabetically/k/korean-war-chronology/may-july-1953.html: accessed 26 September 2020).
[10] Ibid.
[11] Bruhn, *Wooden Ships and Iron Men: The U.S. Navy's Coastal and Motor Minesweepers, 1941-1953*, 225-226.
[12] Inseung Kim dissertation.
[13] "Korean War: Chronology of U.S. Pacific Fleet Operations, May–July 1953"; "HMAS *Culgoa*" (https://www.navy.gov.au/hmas-culgoa: accessed 26 September 2020).
[14] "Korean War: Chronology of U.S. Pacific Fleet Operations, May–July 1953."

[15] "Korean War: Chronology of U.S. Pacific Fleet Operations, May–July 1953"; Inseung Kim dissertation.
[16] Inseung Kim dissertation.
[17] Ibid.
[18] Ibid.
[19] Ibid.

APPENDIX NOTES:
[1] "Patrol Frigate (PF) Index" (http://www.navsource.org/archives/12/08idx.htm: accessed 16 August 2020).
[2] "Ships Sunk and Damaged in Action during the Korean Conflict (Partial)" (https://www.history.navy.mil/research/library/online-reading-room/title-list-alphabetically/s/ships-sunk-and-damaged-in-action-during-the-korean-conflict.html: accessed 19 July 2020).
[3] Montross and Canzona, *U. S. Marine Operations in Korea 1950-1953 Vol. II The Inchon-Seoul Operation*, Appendix E: Task Organization Joint Task Force Seven.

Index

Agar, Augustus Willington Shelton (Cmde, VC, DSO, RN), 183-184
Agar, Rodney (Comdr., RN), 180-181, 183
Akers, Charles O., 255
Alcaraz, Ramon A., 13
Allan, A. N., 159
Allan Jr., Halle C., 18, 52, 256
Almond, Edward M., 50, 52, 69, 82
Andrewes, William Gerrard (Adm. Sir, KBE CB DSOG RN), 19-20, 36-38, 41, 48, 55-56, 61, 66-69, 83, 85, 92, 99, 107, 115, 117, 119, 123, 257
Archer, Stephen Morris, 73-78
Auboyneau, Philippe, 92-93
Australia/Australian
  No. 77 Squadron RAAF base at Kimpo, South Korea, 141
  HMAS Commonwealth (base at Kure, Japan), 168
  Royal Australian Regiment,
    1st Battalion, 135
    3rd Battalion, 224
Barker, F. N., 103
Barnett, Kenneth Malcolm (Comdr., OAM MID RAN), 85
Becher, Otto Humphrey (Rear Adm., DSC, RAN), 56-57, 120, 257
Begg, Varyl Cargill (Adm. of the Fleet Sir, GCB, DSO, DSC, KDtJ, RN), 40
Booth, Blake B., 255
Boyd Jr., Randall T., 77
Briggs, Cameron, 255
Brigham, Carrol W., 255
Britain/British
  Army
  27th British Commonwealth Brigade, 110
  Royal Marines
    41 Independent Commando, Bickleigh Camp, Stonehouse Barracks, 134
  Royal Navy
    Coastal Motor Boat Base, Osea, 183-184
Brock, Jeffery Vanstone (Rear Adm., DSO DSC US Legion of Merit RCN), 45, 56, 81-90, 229, 257
Brock, Patrick Willet (Capt., DSO MID RN), 38, 52, 103, 107, 256
Brothers, J. V. (Lt. Comdr., RN), 55, 256
Burke, Arleigh A., 139, 163
Burke, William A., 116-118
Button, Edward J. (Able Seaman, DSM, RNZN), 135-136
Buzza, Percival Charles (Lt. Comdr., RCN), 195
Cabanie, Urbain E. (Comdr., FN), 55, 256

Caldwell Jr., Sam J., 255
Canada/Canadian
　Esquimalt, 44, 156
　HMCS Stadacona (shore establishment in the Halifax area), 156
　Princess Patricia's Canadian Light Infantry, 110
Castro, Laureano Gomez, 140
China/Chinese
　20th Army, 125
　50th Army, 157
Christiansen, Fran M., 255
Christison, Alexander Frank Philip (Gen. Sir, 4th Baronet, GBE, CB, DSO, MC & Bar, BA), 102
Clark, Harry A., 55, 256
Clark, Jeane R., 255
Clark, Joseph J., 220
Clay, Donald N., 73, 77
Clifford, Eric George Anderson (Vice Adm. Sir, KCB, CBE RN), 224
Close, Robert H., 53, 256
Collier, Andrew Lawrence (Vice Adm., CMM, DSC, CD RCN), 87, 90
Collum Jr., William J., 255
Conrad, Homer E., 55, 256
Converse Jr., Eliot V., 55, 256
Cook, William Henry, 219
Crotty, Bill, 135
Cyr, Joseph C., 155-156
Demara Jr., Ferdinand Waldo, 155-156
Dollard, Allan Nelson (Cdre, DSC RAN), 147-148
Donachie, Frank P. (Lt., RNVR), 102
Dooks, P., 103
Doyle, James H., 49-50, 73, 82, 255
Drysdale, Douglas Burns (Col., DSO, OBE, USA Silver Star, RM), 134
Dukes, Paul Henry, 183-184
Dunn, "Buster" (Stoker Mechanic, RNZN), 136-137
Dyer, George C., 144-147, 165-166
Ereckson, Henry J., 84
Ewen, Edward C., 255
Fuller Jr., Robert C., 55, 256
Gary, Stanley P., 55, 256
Giap, Vo Nguyen, 94
Goodney, William K., 255
Grant, Ferris Nelson (Maj Gen., CB, RM), 134
Greening, Charles Woollven (Capt., DSO, DSN RN), 222
Grkovic, Nicholas, 55, 256
Gromyko, Andrei, 7
Guo, Chai Zeng, 157
Gurnette, Byron L., 256

Hamer, Charles Athelstan (Lt. Comdr., RCN), 163
Harker, Noel (RM), 136-137
Hartman, Charles C., 36
Headland, Edwin H., 52, 256
Henderson, Harry H., 255
Herrick, Laurance Edward (Comdr., DSC RN), 55, 256
Higgins, John M., 20, 36, 52, 256
Hillenkoetter, Roscoe H., 104
Hoare, Peter James Hill (Comdr., OBE, RN), 55, 256
Hodes, Henry I., 139, 163
Hodgson, Derek, 175, 259
Holland, Cedric S., 102
Hone, Hebert R. (Maj. Gen., BA), 102, 287
Hoskins, John M., 255
Howard, Hugh W., 255
Howard, T. R., 55, 256
Hughes, Jack J., 255
Huijeong, Lee, 114
Hunter, John Desmond (Capt., RM), 136-137
Hurl, David William (Lt., RCN), 63-64
Hwa, Chang Chae, 117
Hyoyong, Choi, 102-103
Il, Sohn Won, 11-15, 72, 210
Jackson Jr., Charles B., 255
Jay, Alan David Hastings (Capt., DSO, DSC & Bar RN), 36-37, 43
Jong, Lee Hi, 56, 257
Josephson, Alan R., 55, 256
Joy, Charles Turner, 17-20, 35-37, 43, 68, 72, 82, 99, 105-106, 131, 139, 144, 163,
Kelly, James Maxwell (Lt., DSC RAN), 148
Kelly, Samuel G., 82-90
Kim (ROKN commanding officer), 115-116
King, Dudley Gawen (Capt. DSC, RCN), 161, 185
Kirk, Alan G., 6
Korea
  Geographic locations
    Changnin-do, 179, 190-191
    Changyang-do, 166
    Chinhae, 2, 3, 11, 13
    Chinnampo, 6-7, 37, 39, 51-52, 61-62, 69, 72-92, 99-103, 114, 129, 133, 135, 150-153, 190, 202, 215, 240
    Cho-do, 85-87, 100-101, 114-116, 127-129, 142, 150-166, 176-179, 195-223
    Chonan, 105
    Chongyang-do, 152
    Chulpo, 14, 16, 42

Hachwira-do, 214
Haeju, 44, 51, 62, 67, 103, 107, 120, 132-133, 144-146, 178-223
Hungnam, 69, 82, 104, 149, 202, 243-245
Hwa-do, 149, 246
Hwanghae, 100-108, 113-119, 124-129, 144, 150-158, 176-177, 217
Hyesanjin, 82
Inchon, 13, 16, 35-43, 48-61, 66-67, 72-73, 82, 85, 91-92, 99-114, 120, 126-130, 142-143, 229
Iwon, 42, 73
Ka-do, 159-160
Kaesong, 69, 129-130, 139, 143-146, 174, 207
Kal-to, 179
Kangnung, 8, 126
Kirin-do, 178-179
Kosong, 174
Kusan, 13, 62-63
Losuapto Island, 210
Mahap-do, 179
Masan, 8
Mayang-do, 149
Mokho, 13
Mokpo, 43-44, 51
Mu-do, 186, 188
Nan-do, 150
Ohwa-do, 179, 222
Osan, 8
Paengyong-do, 40, 101, 113-118, 142, 150, 159, 162-166, 177-179, 195, 212, 217-219, 223
Panmunjom, 148, 163, 174, 213, 221-223
Pohang, 13-14, 42
Pusan, 2-20, 35-47, 95, 103-110, 190, 220
Pyongyang, 6-7, 36, 40, 67-72, 84
Samchok, 8, 14
Sanechok, 42
Sangchwira-do, 214
Seoul, 4, 8, 13, 35, 49, 52, 59, 63, 67, 100-104, 125-126, 145, 193
Sochong-do, 101, 217
Sojongjok-to, 162-163
Sok-to, 114-115, 127, 149-166, 176-179, 195, 208-223, 244
Songjin, 141, 149, 243-247
Sosa-ri, 214
Sosuap-to, 179, 187-188, 212
Soyongong-do, 179
Sunchon, 46
Sunwi-do, 179
Suwon, 35, 106

Taechon, 105
Taechong-do/Taechong Islands, 39, 40, 101, 166, 195, 217
Taedong, 84, 86, 128-129, 142, 152, 229
Taegu, 9, 35, 122
Taejon, 35
Taejongjok-to, 162
Taesuap-to, 179, 186, 212
Taewha-do, 159-163
Tan-do, 159-160
Techong-do, 103, 107, 177, 217, 219
Tokchok-do, 104, 177, 195
Ung-do, 152, 166
Wollae-do, 178
Wolmi-do, 52-58, 111
Wolto, 157
Wonju, 104
Wonsan, 37, 42, 51, 61, 69-74, 82, 91-92, 141, 149, 166, 202, 235, 243-247
Yang-do, 141, 149, 220, 247
Yo-do, 220
Yongdok, 42
Yongmae-do, 157, 179, 186-188, 193, 197, 222
Yongpyong-do, 165, 177-179, 186-188, 195, 217, 219
Yosu, 45-47
Yuksom, 179
Yuk-to, 178
Military
  Democratic People's Republic of Korea (North Korea)
    3rd Army Division, 4th Army Division, 8
  North Korea anti-Communist guerillas
    Pyeongyang Partisan Regiment, 101
  Republic of Korea (South Korea)
    1st Division, 67
    1st Korean Marine Corps Regiment, 59
    I Corps, 118, 126
    II Corps, 82
    III Corps, 126
    6th Infantry Division, 125
Laird Jr., William R., 256
Landymore, William Moss (Rear Adm., OBE MID(2) RCN), 194, 196
Lee (Maj. Gen., ROKA), 163
Lee, Herbert Jack (Capt., CBE, DSC, RN), 56, 257
Lesh, Rohan Edwin (Comdr., RAN), 85
Levin, Philip, 55, 256
Libby, Ruthven E., 163
Lilly Jr., Percy A., 55, 256
Lloyd-Davies, Cromwell Felix Justin (Capt., DSO, DSC, RN), 69, 99, 102-

103, 114-115, 142, 162
Lonvik (RCN sailor), 172
Lopez, Baldomero, 59
Lowry, W. M., 255
Lungren, Oscar B., 53, 256
MacArthur, Douglas, 8, 17, 45, 49-52, 67-70, 82, 104-107, 124, 140
Mahoney, Terry, 224
Malenkov, Georgy, 212
Malik, Jacob A., 139
MacFarlane, George Richard (Comdr., RCN), 193-196
Mansergh, Eric Carden Robert (Gen., Sir GCB, KBE, MC BA), 102
Marchioni, Robert (Able Seaman, RNZN), 136-138
Marks, William Beresford Moffitt (Capt., CBE, DSC, US Legion of Merit, RAN), 56, 87, 229, 257
Martin, Harold M., 124
McCabe, George, 11
McConechy, Earl Edward (Lt. Comdr., RCN), 195
McDonald, Ian Hunter (Capt., RAN), 18
McGee, John, 108, 116-117
McGregor (Corporal, RM), 136-137
McKechnie, Arnold W., 256
McKinnon, A. L., 120
McLachlan, M., 120
McMahon, James P., 55, 256
Meyer, William C., 256
Moore, Elmer, 255
Muccio, John J., 7, 35
Nam, Chai Yong, 56, 257
Nichols, Donald, 152
Ogle, Elvin C., 255
Operation
  LITTLE SWITCH, 213
  PANDORA, 217-223
  RUGGED, 124
  SHINING MOON, 117
  SICIRO, 192-198
Parker, Charles W., 255
Peacock, Thomas Stewart Reid (Comdr., RCN), 63
Pearson, Douglas James (CPO, BEM RCN), 90
Philippines
  Military
    2nd, 10th, 14th, 19th, 20th Battalion Combat Teams, 13
Phillips, Charles E., 255
Plomer, James (Cmde, OBE DSC & bar, RCN), 160
Quirino, Elpidio Rivera, 13
Radel, Frederick M., 53, 256

Index   291

Reiter Jr., Harry L., 255
Revers, George, 94
Reyes, Julio Cesar, 140
Rhee, Syngman, 220-221
Rice, Lester K., 255
Ridgway, Matthew B., 104, 106, 111, 124, 148
Ripley, Richard M., 192
Rochotte, Leon C., 91
Roeder, Bernard F., 255
Romer, Robert D., 74
Ruble, Richard W., 256
Salter, Jocelyn Stuart Cambridge (Vice Adm., DSO, OBE, US Bronze Star, RN), 38, 41, 52, 256
Saunders, Willard A., 93
Schelling, Robert A., 53, 256
Scoles, Norman J. (Leading Seaman, DSM, RNZN), 135-137
Scott-Moncrieff, Alan Kenneth (Adm. Sir, KCB, CBE, DSO & Bar RN), 119-120, 123, 127-128, 147, 152, 158, 164, 173, 178
Seibert, Donald A., 121
Self, Cecil (RM), 136
Shields, Thomas, 64

Ships and Craft
  Australian
    Anzac, 208-209, 251-253
    Bataan, 20, 24, 34, 37, 43-44, 56, 65-69, 85-89, 100, 104, 118, 192-193, 225, 229, 249, 252-257
    Condamine, 24, 29, 172-173, 212, 252
    Culgoa, 24, 221-222, 252-253
    Murchison, 24, 97, 145-148, 228, 250-251
    Shoalhaven, 18-24, 34, 37, 41-43, 249-250
    Sydney, 24, 62, 154, 160, 162, 180, 226, 231
    Tobruk, 24, 166, 251, 253
    Warramunga, 24, 34, 56
  British
    Alacrity, 20, 22, 34, 38, 166, 249-251
    Alert, 22, 34, 126, 130, 137, 142, 161, 172, 223, 250-251
    Amethyst, 22, 115, 120, 147, 185, 190-191, 250-252
    Belfast, 20-23, 34-38, 44, 109, 115-120, 155-162, 190-195, 249-251
    Birmingham, 22, 219-223, 252-253
    Black Swan, 14, 20, 22, 34-38, 43, 120, 147, 249-251
    Brown Ranger, 45
    Cardigan Bay, 22, 34, 69, 100, 142-147, 250-252
    Ceylon, 22, 34, 55-56, 62, 67, 69, 83, 86, 100-109, 115, 136, 141-142, 149, 156, 162-166, 187, 228, 250, 252, 257
    Charity, 22-23, 34, 37, 44, 52-57, 249-257
    CMB-4, 184

Cockade, 22, 34-38, 44, 56-57, 67, 109, 160-162, 249-257
Comus, 22, 34, 36, 48, 120, 186-188, 208, 249-252
Concord, 22, 34, 57, 100, 175, 180-181, 250-253
Consort, 20, 22, 34-38, 115, 249-252
Constant, 34, 250-252
Constance, 22, 69, 99-100, 166
Cossack, 20-22, 34-44, 69, 100, 194-198, 249-253
Crane, 22, 251-253
Glory, 22, 62, 127, 129, 142-144, 169
Hart, 20, 22, 34, 37, 43, 249-250
Jamaica, 14, 20, 22, 34-38, 52, 62, 249-250, 256
Kenya, 22, 34-38, 52, 69, 100-107, 116, 141, 249-251, 256
Modeste, 22, 219, 252-253
Morecambe Bay, 22, 34, 69, 100, 145-147, 210, 215, 250-253
Mounts Bay, 22, 34, 43-47, 55, 67, 91-92, 147, 250-256
Newcastle, 22, 215-219, 252-253
Ocean, 22, 68, 198, 223-224
Opossum, 22, 252
Sparrow, 22, 252-253
St. Brides Bay, 22, 34, 147, 198, 214-215, 250-253
Sunflower, 160
Telemachus, Tyne, Unicorn, 22
Theseus, 22, 62, 67, 74-77, 83-84, 99-104, 119
Triumph, 19-22, 35, 55-57, 62, 67, 257
Whitesand Bay, 22, 34, 55, 91, 161, 250-256
Canadian
  Athabaskan, 26, 34, 37, 44-47, 56, 62-69, 85-89, 100, 158-163, 167-171, 176, 185-189, 229, 249-257
  Cayuga, 26, 34, 37, 43-47, 56, 65, 69, 81-90, 100, 109-112, 154-160, 229, 249-257
  Crusader, 26, 208-209, 252-253
  Haida, 26, 214, 252-253
  Huron, 26, 250-253
  Iroquois, 26, 192-198, 223, 252-253
  Nootka, 26, 109-111, 120, 195-210, 250-252
  Sioux, 26, 34, 37, 44-47, 56, 69, 85-87, 100, 141, 224, 229, 249-257
Colombian
  Almirante Brion, Capitan Tono, 30, 241-242, 252-253
  Almirante Padilla, 30, 96-97, 140-142, 251
Danish, *Hertmaersk*, 95
Dutch
  Evertsen, 29, 34, 37, 47, 56, 69, 100, 212, 249-251, 257
  Johan Maurits van Nassau, 29, 212
  Piet Hein, 29, 212, 251-252
  Van Galen, 29, 128-129, 212, 250-251
French

La Grandiere, 31-37, 43, 55, 67, 249-250, 256
German
 U-631, *U-638*, 160
Japanese
 MS-03, *MS-06*, *MS-08*, *MS-09*, *MS-10*, *MS-12*, *MS-13*, *MS-15*, *MS-21*, *MS-22*, *MS-23*, *MS-57*, *PS-56*, *PS-62*, 74
 LST-Q007, 74
New Zealand
 Hawea, 25, 131, 133, 147, 250-253
 Kaniere, 25, 131, 251-253
 Pukaki, 25, 34, 37, 47, 55, 67, 91, 131-134, 249-250, 256
 Rotoiti, 25, 34, 69, 100, 131-138, 147, 250-252
 Taupo, 25, 131, 147, 251-252
 Tutira, 25, 34, 37, 47, 55, 67, 69, 91, 100, 131-133, 249-250, 256
Norwegian, *Reinholt*, 35
Democratic People's Republic of Korea (North Korea) *Kimball R. Smith*, 5-6
Philippine, *Albay, Bulacan, Capiz, Cotabato, Misamis Oriental, Negros Oriental, Pampanga*, 13
Republic of Korea (South Korea)
 Apnokkang, 115, 124-125, 145, 233
 Bak Du San, 1-9, 48, 188, 233
 Chiri San, 15-16, 56, 63, 86, 107, 115, 233, 257
 Chungmugong (I), 12, 214, 234
 Chungmugong (II), 234
 Daedonggang, 12, 235
 Daegu, 12, 56, 86, 89, 114, 235, 257
 Daejeon, 12, 89, 188, 235
 Dancheon, 12, 56, 235
 Danyang, 12, 56, 63, 74, 235, 257
 Deokcheon, Dumangang, Gapyeong, Gowan, Gyeongsan, Tongcheon, 12, 234-235
 Dumon, 102, 233
 Gaeseong, 12, 48, 234
 Galmaegi, Gireogo, 210-212, 233
 Ganggye, 12, 214, 234
 Ganggyeong, 12, 56, 234, 257
 Geum Seong, Han Ra San, Heukjohwan, Jebi, Mok Seong, 233
 Gongju, 12, 70, 234
 Guwolsan, 12, 48, 56, 132-133, 234, 257
 Hwa Seong, 210, 233
 Imchin, 168, 233, 241
 Kang Jim, 12, 56, 234, 257
 Kang Nung, Kang Wha, Kaya San, Kim Hae, 12, 234
 Kil Chu, 12, 48, 211, 214, 234
 Kim Chon, 12-16, 42, 74, 234
 Ko Yung, 56, 132-133, 234-235, 257
 Kum Kang San, 15-16, 42, 48, 56, 191, 233, 257

294   Index

  Kwang Chu, 12, 43, 48, 56, 74, 77, 132-133, 234, 257
  Kyong Chu, 12, 43, 56, 74, 133, 234, 257
  Myo Hang San, Olbbaemi, Pongnoe, 233
  Nae Tong, 233, 241
  Paik Doo San, 56, 257
  Sam Kak San, 15-16, 42, 51, 56, 233
  Su Seong, 214, 233
  Taebaeksan, 12, 114, 235
  Taedong, 97, 233, 241
  Tongpliu, 94
  Tongyeong, 12, 56, 89, 235, 257
  Toseong, 12, 86, 235
  Yong Kung, 12, 56, 235, 257
 Russian, *Andrei Pervozvanny, Oleg, Pamiat Azova, Petropavlovsk*, 183-184
 Thai
  Bangpakong, 27-28, 34, 94-95, 250-251
  Prasae (I), 27, 95-96, 250
  Prasae (II), 27-28, 34, 242, 251, 253
  Tachin, 27-28, 241, 251, 253
 United States
  Merchant Marine *Ensign Whitehead*, 2-3
  Navy
   amphibious/transports
    Algol, Montague, Okanogan, 82
    Bayfield, 82-83, 87
    Bexar, 82, 88
    Catamount, 74, 76
    El Dorado, 117
    Epping Forest, 143
    General C. G. Morton, 45
    Horace A. Bass, 74, 79, 105, 240
    LSMR-412, 208
    LST-529, 220
    Mount McKinley, 50
    Sgt. Sylvester J. Antolak, 13
    Whetstone, 142-143, 152
   auxiliaries and service vessels
    Ajax, 168
    Alchiba, Algorab, Aquarius, Centaurus, Cepheus, Serpens, 5-6
    Bolster, 74
    Bryce Canyon, 78
    Dixie, 72, 92
    Jason, 97
    Repose, 77
   combatants
    aircraft carriers

Badoeng Strait, 62, 104, 106, 180, 188, 256
Barioko, 62
Bataan, 23, 62, 106, 115, 119, 128-129, 142, 188-191
Boxer, 42, 68, 255
Philippine Sea, 255
Rendova, 62, 163
Sicily, 62, 104, 106, 142-144, 192-198, 256
Valley Forge, 35, 68, 255
battleships
Missouri, 17, 113, 220, 237
New Jersey, 237, 243
cruisers
Los Angeles, 146, 237, 246
Rochester, 91, 104-106, 237, 256
Toledo, 42, 52, 60, 123-130, 237, 256
Worcester, 237, 255
destroyers
Chevalier, Fletcher, The Sullivans, 238, 255
Collett, 18, 41-41, 53,54, 238, 243, 256
De Haven, 18, 35, 41-42, 53, 256
Ernest G. Small, 238, 244, 256
Eversole, 166, 238, 255
Forrest Royal, 74, 76, 86-89, 229, 238
Frank Knox, Hamner, Hollister, 239, 255
George K. MacKenzie, 237, 256
Gregory, 71, 238
Gurke, 53, 237, 239, 243, 247, 256
Hanson, Higbee, James E. Kyes, 247, 255-256
Henderson, 53-54, 237, 239, 244, 247, 255
Lewis Hancock, 168, 239
Lyman K. Swenson, 18, 41-42, 53-54, 60, 209, 237, 239, 243, 256
Mansfield, 18, 35, 41-42, 52-53, 237, 243, 246, 256
McGowan, 168, 237
McKean, 52, 237, 255
Ozborne, 255
Power, 73
Radford, 237, 255
Rowan, 55, 238, 244, 247, 256
Shelton, 238, 244, 255
Southerland, 55, 239, 245, 256
Taussig, 239, 246, 256
Theodore E. Chandler, 238, 255
Walker, 239
Wiltsie, 239, 247, 255
patrol frigates
Bayonne, Evansville, Newport, 55, 240-242, 256

mine warfare
   minesweepers
      Carmick, 74-79, 105, 240
      Chatterer, 18, 55, 58, 67, 72, 256
      Condor, 209
      Doyle, 73, 240
      Endicott, 73, 96, 240-245
      Gull, 74, 79, 246
      Incredible, 18, 72
      Kite, 18, 55, 58, 67, 72, 246, 256
      Mainstay, 18
      Merganser, 58, 72
      Mockingbird, 18, 55, 58, 66-67, 72, 256
      Osprey, 18, 55, 58, 72, 244-246, 256
      Partridge, 18, 55, 58, 72, 243, 256
      Pelican, 74, 79, 211
      Pirate, 18, 70-73, 243
      Pledge, 18, 55, 58, 67, 70-73, 243, 256
      Redhead, 18, 55, 58, 67, 72, 256
      Swallow, 74-79, 105, 245
      Thompson, 74-79, 240-246
      YMS-413, 132
   seaplane tender *Gardiners Bay*, 74, 77

Shouldice, Darcy V., 18
Shumer, Ed, 42
Sibak, Hyun, 56, 257
Skelton, Clifford, 161
Skipper, Jack Harold (Lt., MC, AA), 135
Smith, Allan E., 72-73, 83, 86, 92, 106, 115, 123, 128, 141, 144
Smith-Cumming, George Mansfield (Sir, KCMG, CB), 183
Snow, Franklin C., 256
So, Lee Hung, 51, 56
Sok, No Kum, 141
Sowell, Jesse C., 18
Spofford, Richard T., 55, 67, 91, 256
St. Clair-Ford, Aubrey (Capt. Sir, BT DSO RN), 115-117, 159
Stalin, Joseph, 212
Stewart, William S., 53, 256
Stone, James Riley (Col., DSO & Bar CA), 110
Stout, Richard F., 52, 256
Struble, Arthur D., 17, 19, 91-92, 104, 124, 255
Stuart, Archibald W., 218
Sturgeon, Clarence Alfred (Lt. Comdr., RCN), 65
Sung, Kim Il, 212
Swenson Jr., David H., 60
Smyth, Dacre Henry Deudraeth (Cdre, AO RAN), 225-232

Taplett, Robert D., 7
Taylor, Harry (Padre, RNZN), 138
Taylor, Paul Dalrymple (Capt., RCN), 45, 56, 257
Thach, John S., 256
Thackrey, Lyman Augustus, 82-83, 90, 104-105, 126, 130
Thompson (Col.), 117
Torlesse, Arthur David (Rear Adm., DSO RN), 56, 257
Truman, Harry S., 8, 33, 104, 140
Turner Brian Edmund (Cdre, OBE, DSC, RNZN), 132-138
Turner, Howard, 163
United States
  Air Force
    Fifth Air Force,
      22nd Crash Rescue Boat Squadron, 151
      6004th Air Intelligence Service Squadron, Detachment 2, 151-152
    Combined Command Reconnaissance Activities, Korea, 122, 152, 218
  Army
    2nd Infantry Division, 8, 82
    3rd Infantry Division, 150
    7th Infantry Division, 52, 82
    7th Medium Port and 501st Harbor Craft Platoon, 84
    Tenth Corps, 50, 52, 69-70, 82, 104
    24th Infantry Division, 8, 125
    25th Infantry Division, 8, 125, 150
    442nd Military Intelligence Detachment (codename Salamander), 150
    8240th Army Unit, 122
    Camp Wood, 167-172
    Leopard Force, 108, 116-127, 142, 150, 157, 159, 166, 177, 180
    Task Force Kirkland, 150
    Wolfpack, 188, 192-198
  Marines
    1st Marine Division, 52, 58, 70, 82, 125, 134
    1st Provisional Marine Brigade, 7-8
    5th Marine Division, 58
  Navy
    Destroyer
      Division 31, 255
      Division 32, 129
      Division 91, 18, 42
      Destroyer Division 111, 112, 255
      Flotilla Five, 78
      Squadron Thirty, 93
    Cruiser
      Division 1, 104
      Division 5, 139
    Helicopter Utility Squadron One (HU-1), 74

Mine Squadron Three, 57
  Mine Division 4, 73
  Mine Division 31, 18, 58
  Mine Division 32, 18
 Minesweeping Boat Unit, 79
 Transport Squadron One, 82, 90
 Underwater Demolition Team, 104-105, 147
  One, 74, 79
  Five, 202
Unwin, John Henry (Rear Adm., CB, DSC, RN), 43, 55, 256
Van Doorninck, D. J. (Lt. Comdr., RNN), 56, 257
Van Fleet, James, 124-126
Vliet, C. van, 212
Walker, Walton, 8, 69, 82, 104
Webber, Richard Scott Foster (Lt., DSC, RN), 136
Welch, David Fife 'Kelly,' 79
Welland, Robert Phillip (Rear Adm., DSC RCN), 45, 56, 257
Welte, Cecil R., 256
Willis, Albert C., 6-7
Wilson (Col.), 88
Wingrove, Norman D. (Brig. Gen., BA), 102
Woodyard, Edward L., 52, 256
Worth, Peter Reginald Glenholme (Comdr., DSC RN), 56, 257
Wright, Edwin K., 50
Wright, Erma A., 93
Yong, Nam Choi, 5
Young, David (Lance Corporal, AA), 135
Young, Lloyd V., 18
Young, Richard, 55, 256
Yule-jung, Kwon, 9
Zedong, Mao, 212

About the Author

Commander David D. Bruhn, U.S. Navy (Retired) served twenty-two years on active duty and two in the Naval Reserve, as both an enlisted man and as an officer, between 1977 and 2001.

After completion of basic training, he served as a sonar technician aboard USS *Miller* (FF-1091) and USS *Leftwich* (DD-984). He was commissioned in 1983 following graduation from California State University at Chico. His initial assignment was to USS *Excel* (MSO-439), serving as supply officer, damage control assistant, and chief engineer. He then served in USS *Thach* (FFG-43) as chief engineer and Destroyer Squadron Thirteen as material officer.

After graduation from the Naval Postgraduate School, Commander Bruhn was assigned to Secretary of the Navy and Chief of Naval Operations staffs as a budget analyst and resources planner before attending the Naval War College in 1996, following which he commanded the mine countermeasures ships USS *Gladiator* (MCM-11) and USS *Dextrous* (MCM-13) in the Persian Gulf.

Commander Bruhn's final assignment was executive assistant to a senior (SES 4) government service executive at the Ballistic Missile Defense Organization in Washington, D.C.

Following military service, he was a high school teacher and track coach for ten years, and is now a USA Track & Field official. He lives in northern California with his wife Nancy and has two grown sons, David and Michael.

www.ingramcontent.com/pod-product-compliance
Lightning Source LLC
Chambersburg PA
CBHW071954220426
43662CB00009B/1126